UNDERGRADUATE EDUCATION IN THE SCIENCES FOR STUDENTS IN AGRICULTURE AND NATURAL RESOURCES

SUMMARY OF PROCEEDINGS OF REGIONAL CONFERENCES

Commission on Education in
Agriculture and Natural Resources

NATIONAL ACADEMY OF SCIENCES

WASHINGTON, D.C. 1971

The work of the Commission on Education in Agriculture and Natural Resources was supported through a contract between the National Academy of Sciences and the National Science Foundation.

ISBN 0-309-01921-4

Available from

Printing and Publishing Office
National Academy of Sciences
2101 Constitution Avenue
Washington, D.C. 20418

Library of Congress Catalog Card Number 77-169177

Printed in the United States of America

Preface

In November 1966 the Commission on Education in Agriculture and Natural Resources, with the Commission on Undergraduate Education in the Biological Sciences, sponsored a national conference under the title "Undergraduate Education in the Biological Sciences for Students in Agriculture and Natural Resources."* It was cosponsored also by the Resident Instruction Section, Division of Agriculture, National Association of State Universities and Land-Grant Colleges and by the National Association of Colleges and Teachers of Agriculture.

As a follow-up to this conference, four regional conferences were held in the ensuing few years, bringing together teaching faculty members from agriculture, forestry and other natural resource areas, and biology. While the participants were different at each conference and reflected slightly different emphases, they were individuals considered to have appreciable influence upon undergraduate programs in their respective institutions and included the chairmen or members of college or department curriculum committees.

*Commission on Education in Agriculture and Natural Resources. 1967. Undergraduate Education in the Biological Sciences for Students in Agriculture and Natural Resources. Publ. 1495, National Academy of Sciences, Washington, D.C. 86 p.

The Commission recognized two general objectives of the regional conferences: (1) to acquaint the instructional staffs in agriculture and natural resources with the changes that are taking place in their areas of concern and with the opinions of various study groups as to the education in the sciences needed by their students over the next few decades, and (2) to inform biologists of the needs of students in agriculture and natural resources, for their consideration in planning core curricula and service courses. There were four conferences in all:

- The Western Region Conference was held at the University of Nevada, Reno, February 22–23, 1968, under the chairmanship of Darrel S. Metcalfe, University of Arizona.
- The Northeast Region Conference was held at the Hotel America, Hartford, Connecticut, May 2–3, 1968, under the chairmanship of Russell E. Larson, Pennsylvania State University.
- The North Central Region Conference was held at the University of Wisconsin on March 13–14, 1969, under the chairmanship of Carroll V. Hess, Kansas State University.
- The Southern Region Conference was held at the University of Georgia on October 21–22, 1969, under the chairmanship of Hal B. Barker, Louisiana Tech University.

The general similarity of the regional conferences suggests that the most practical way to organize the proceedings is to consolidate like topics within a given chapter and to maintain the integrity of the individual presentations as essays within that chapter. This approach has the virtue of eliminating a certain amount of duplication and making all the various viewpoints conveniently available. Where two speakers on the same general topic approach it from different viewpoints or take opposite points of view on particular facets, both statements have been retained essentially as initially presented.

The Commission is grateful to the very substantial number of persons who participated in organizing the regional conferences, appeared as invited speakers on the platform, or joined the informal discussions as attending invitees.

Contents

1

Trends in Agricultural Curricula

R. L. KOHLS

I propose to discuss some guidelines in curricula improvement from the viewpoint of a professor who has been teaching undergraduates for nearly 20 years, who has had charge of departmental faculty curriculum development activities for many of these years, who has had a brief exposure to departmental administration and who now finds himself viewing the university as assistant to the academic vice president.

We must cease "running scared" in agriculture. A primary handicap to constructive thought and change is the continuing debate about whether or not the school of agriculture—call it by any name you choose—has an educational role to play in the future. Change is not unique to agriculture. Anyone who observes the university scene today sees the challenge of change in almost every area. Questions are being asked in almost all the professional schools—home economics, engineering, and pharmacy, to name several—as to roles and curricula. Institutions do not die unless they refuse to adapt, and I strongly urge that we accept the proposition that the better agricul-

1

tural schools are going to be around for many years. Such an attitude
will be like a breath of spring air to faculties and may stimulate
growth and renewal.

Part of the problem is uncertainty whether the agricultural school
is merely an alternative route to a good general or scientific education or a substantial, mission-oriented school serving professional
needs in a dynamic society. If one accepts the first view many of the
doubts we hear are very real. Personally, I prefer the concept of a
professional school that offers training in applied disciplines to serve
a particular set of educational needs. This position, taken with pride
and firmness, is one of the first steps that can combat the inferiority
complex plaguing many of our faculty.

Another difficulty is that we often have not grown along with the
professional needs of our students. It has been proclaimed for years
that the professional fields served by our school are appropriately
those that deal with the total food and fiber complex—from the farm
through to the consumer. "Agribusiness" was a term coined to cover
this complex, though I prefer "agindustrial science and technology."
To this we have recently added a concern for the development and
quality of rural life.

If these are the true professional concerns of our students, then
people and personnel management are as important as cattle, corn
and cotton. Too often, however, the primary emphasis continues to
be on the production sciences, whereas food processing, food development and distribution, and the social aspects of community development have been accorded a lesser priority. If we really accept a
broadened definition of our professional mission, then we need put
our educational money where our leaders' mouths have been.

GUIDELINES AND ISSUES

● All professionals are members of society and their higher education should equip them for leadership within it.

● A professional of tomorrow must have a working understanding
of science and mathematics related to his specialty and should be
equipped to think through problems that confront him—he should
not merely be skilled in doing the tasks of the moment.

● The ability to deal with complex systems and problems as a
whole must be emphasized, for really important questions do not
come in specialized bits and pieces.

● Our curricula need expanded opportunities for interdisciplinary work. Traditional departments, though they furnish a very important administrative and educational focal point, can be deadening if they insist that they alone have the answer to all problems. Much of education today requires the joint effort of many different types of professionals, just as in research we create various institutes and centers to get the job done.

● Total credit hour requirements need to be reduced—simply doing more and more of what we are currently not doing especially well is not the route to improvement. In many instances, students take well over four years to complete their curricula.

● We must continually clarify and unify basic knowledge as the flow of new developments continues. Yesterday's principles and doctrine are not necessarily the best. On the other hand, how close to the frontier of knowledge should the basic undergraduate presentation of fundamentals be? On this issue, of course, is focused much of the present concern with the content of biology.

● Courses in the curricula often have too many prerequisites. While sequential education is a proper goal of curricular development, prerequisites too often serve as administrative devices for restricting students rather than as educational stepping stones.

● More stimulating introductory courses are needed to appeal to our educationally improved freshmen; on the urgency and importance of this issue, there is no question.

DIVERSITY IN CURRICULA

Curriculum development should be undertaken in the context of the organization and the educational goals of a given institution—universities and colleges differ widely as to student mix and educational goals. There is a great tendency toward mimicry among education institutions, a great urge to find out what some "leading" institutions are doing and then try to duplicate it. But in one institution it may make sense for the agricultural school to teach its own basic biology, biochemistry, or statistics, whereas in another to do this would duplicate educational effort. We should accept our institution and its purposes for what they are and build on these in developing curricula.

Curricula must take into account the differences among student bodies and among educational goals. Too often, there is a tendency to make all undergraduate programs simply preparatory to graduate

training and to regard only "hard" science curricula as "first class"—
an unforgivable intellectual snobbery. In the same way it is false to
maintain that good undergraduate education in agriculture cannot be
provided in some of our smaller new colleges because they do not
have the research farms, laboratories, and other resources of a large
university.

There is, of course, a tremendous personal satisfaction to a professor to see his students go on to graduate work, but I submit it does
not necessarily satisfy the educational goals of all students. The real
challenge is to make each curriculum of high quality, pertinent, and
well taught. It is sobering to remind ourselves that the vast majority
of those who will manage our farms and businesses and who will
govern our society do so without the benefit of graduate education.

CORE REQUIREMENTS

Careful attention should be given to what is desired in the core educational requirement for a school and to its general educational content. The world "core" implies minimal necessary requirements and
should not be confused, as faculties often do, with the concept of
optimum exposure. Because each specialized group wants the student
to have optimal exposure in its particular area, too many efforts at
curricular revision end up with more required courses and increased
total credit hours.

In deciding on core and content we must face squarely the great
professional diversity within a school. It makes as little sense to require business majors to take chemistry or biology in depth as it does
to require soil science or nutritional majors to take advanced theoretical economics. If the school is to have a single core curriculum for
all, the hard question of minimum must be faced.

The concept of a single minimum core forces us to decide what we
want to accomplish in the "general education" part of the curriculum.
Too often the "general education" needs are met by requiring that
students take some English, literature, history, sociology, and economics. Probably we do somewhat better with the core requirements
in the sciences as a way of building background in professional
specialities.

My experience is that the core and general education requirements
of the first two years cause much student anguish and frustration.
They continually ask "why?" and "what for?"—questions the faculty

find hard to answer because they are not sure themselves.

More dialogue is needed between the agricultural professional school and the rest of the university. We have been indiscriminate users of the output of others in the academic community and have had little to say about the designs of the general education offerings we use. But we cannot have much input if we cannot first formulate our own needs, problems, and goals.

The rest of the university has the same "we-would rather-teach-only-our-own-majors" view as do our own departments—a service role is enthusiastically performed at very few places on the university campus. Certainly we cannot settle for watered down, old-fashioned, out-of-date offerings for servicing our students.

CURRICULA CHOICE

It is a question how many curricula options there should be, considering the present diverse professional emphases of our schools. I doubt that there is any single answer. If we insist upon tight, highly prescribed curricula, then we must have several routes if we are to satisfy the diverse student needs and goals. If we have a more loosely prescribed curriculum and a substantial number of electives, then we can have fewer basic routes.

Loose, highly elective curricula cannot be automatically equated with excellence. Neither the straitjacket nor chaos has much to offer the educational system. Tightly controlled and course-specified curricula require continual academic policing to see that the courses are doing their prescribed jobs. Permissive curricula require a very high input of knowledgable advisers and counselors if students are to reach respectable educational goals. I have an uneasy feeling that the present faculty attitude favoring more freedom is often based on a competitive effort to woo students rather than a carefully thought out objective.

ACCOMPLISHING CURRICULA REFORM AND REVISION

Curriculum building is faculty business—it is axiomatic that no worthwhile eudcational operation can be imposed from above. If the faculty is not involved, interested, and desirous of curricula reform, anything that is done will be half-hearted.

A professor's first loyalty is to his professional peers, as a member of the profession into which he aspires to bring his students. He is probably next most concerned about his own course and students, about his department, and about the students majoring therein. Still further down the scale is his concern for the school, the university, and the student's total educational experience.

What I am suggesting is that the very individual who must be the most directly involved in successful curricula improvement is, if left to his own devices, the least interested and least well equipped to do it. This implies that in any curricula effort that is focused on departments, the professors' individual interests will come first, the specialty program for its student majors next, and the total curricula organization of the students and of the school will be a poor third. The good administrator recognizes that he must stimulate and involve professors. The dean and his administrative colleagues should be the best informed educators on the staff, whereas the individual professor is expert in his specialty.

Curricula building must be experimental. Established procedures and ideas need to be challenged, alternatives proposed, put into practice and evaluated. For example, are we overburdened by too many or useless prerequisites? If so, let's do away with some of them and see whether the educational result is downgraded. Are we boxed in by the three credit-hour tradition? Let's experiment with five- or six-hour complexes to handle sequential areas; or one- and two-hour seminars to deal with more specialized, limited problems. This may short-circuit the computer scheduling, but the computer should serve the educational venture, not dominate it.

Do we really need laboratories in introductory chemistry or physics? Let's try to do without it and evaluate the results.

Do we really believe that we should take advantage of the student's high school training and permit him to move on from that point? Advanced placement is used some in mathematics and chemistry. Similarly, high school students who come through a vocational agricultural program should probably be permitted to skip many of the freshman agricultural survey courses.

Would the use of outside consultants aid curricula review and revision? They are commonly used in reviewing research programs. And faculty members believe strongly in their effectiveness as outside consultants to businesses, holding that an informed outsider can more easily see important problems and more freely raise pertinent questions. If the consultant idea is valid elsewhere, why shouldn't consultants be useful in evaluating curricula proposals?

Lewis Mayhew* points out that because curricula revision involves vested interest, self study inevitably becomes conservative political action. You may have heard the statement that changing a college curriculum is like trying to move a cemetery, the problem is that the dead have so many friends.

A principal purpose of good administration is to stimulate and guide the faculty in such a way that it is forced to carry out its academic responsibilities. The administration must create the atmosphere that encourages analysis, experimentation, and change, which goes far beyond periodic, perfunctory "reviews of the curricula," introducing new course titles for old course content, and placing old courses in a new sequence.

The principal tool available to the administrator is the committee. Though committees are the favorite faculty synonym for useless activity, they are the *sine qua non* of college and university decision-making. A good administrator gives critical thought to committee structure, sees that resources are made available for its work, and shows continuing concern over its progress. Most importantly, he makes certain that the faculty sees action taken on the basis of committee recommendations. Much of the cynicism and disillusionment arises from the feeling on the part of faculty members that the administrator really is not interested and does nothing to implement their advice. Implementation can, of course, be either positive action or explicit decision not to act. I think most professors would plead with their administrators not to get so bogged down in keeping the books, checking on vacation plans, and seeing that students are properly registered and grades are properly recorded; they cannot give time and effort to what should be their other important role—the academic leadership of the faculty. This, in the final analysis, is the key to effective curricula revision and improvement.

L. C. PEIRCE

We are a people living within constraints—constraints imposed by federal laws, state regulations, local ordinances, university rules. Such

*L. B. Mayhew. 1967. The Collegiate Curriculum—An Approach to Analysis, Research Monograph No. 11, Southern Regional Educational Board, Atlanta, Ga.

is the daily confrontation with these constraints that we tend to forget that we still have the freedom to think and to act. This freedom is not unlike that expressed in historian Bruce Catton's analysis of the pioneer spirit in a vibrant U.S. republic of the early 19th century.*

The people could go anywhere they chose, quite literally anywhere; all the way to the undiscovered mountains and the deserts, beyond these to the extreme limit of the imagination. Men could very likely do anything on earth they had the courage to dream of doing.

I quote this passage as a good philosophy for any dealing with education and curricula. Credits, quarters, textbooks, labs, the curriculum itself and the prerequisites entrenched therein often represent a straightjacket constraining the development of the new and the bold. Established educational patterns must not encumber deliberations about agricultural curricula and the needs of tomorrow.

But to confront the needs of tomorrow, we must plant our feet—survey where we started, where we are, and, most important, why.

TRENDS IN AGRICULTURE

I have chosen two graphs that together reflect conglomerate progress in agriculture. Figure 1 records the number of horses and mules on

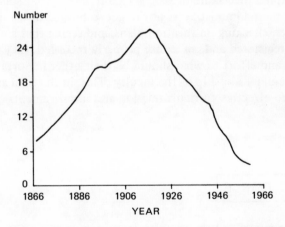

FIGURE 1   Number of horses and mules on United States Farms for the years 1866–1966.

*Agriculture/2000. 1967. (A collection of essays by Orville Freeman, Secretary of Agriculture.) USDA.

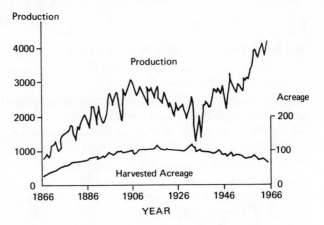

FIGURE 2 Total production and harvested acreage of
corn in the United States for the years 1866–1966. (From
USDA Statistical Reporting Service Agr. Handbook 318.)

farms over a period of 100 years. Figure 2 presents yield and harvested acreage for corn over a similar period of time. These graphs
reflect the single outstanding trend in American Agriculture—efficiency: efficiency of motion, of time, of land and of the inherent
adaptation of the plant or animal to its environment.

Increasing efficiency in agriculture has, since World War II, provided a challenging paradox. The first disturbing question was not
"how to produce more," but "how to produce less." Overabundance
brought overconfidence, a take-it-for-granted attitude that current
warning signals have not yet tempered. It is hard to worry—to be
really concerned—about food supplies when one is full of prepeeled,
precooked victuals bathed in seasoned butter sauce, served with a
freeze-dried salad in a preformed aluminum tray that, when washed,
can serve as a convenient depository for pins, buttons, nuts, and bolts.
This just has to be utopia.

But the agricultural specialist can read the signals. The elevators
are empty of grain; land is disappearing under concrete, asphalt, and
two-car garages. The problem is that the after-effects of the abundant-
era policies still exist. We are now told that before the century is over,
the world food problem will demand the best minds, that agriculture
will become the NASA of tomorrow. In short, money and men, the
strengths of colleges of agriculture in postwar years, will return. But
even if this does come to pass, it is no license for inaction now. And
whether we enter a new golden age or not, it will not be the same.

TRENDS IN CURRICULA

The major changes in agricultural curricula have been from hands to mind, from vocational to professional, from facts to concepts. The end product of our curriculum is not expected to be conversant in the classics, but is now visualized as a humanized computer—able to digest various inputs and to spew out a rational, well-based output. If we ask the graduate to be a computer, then teachers must assume responsibility for programming him. And just as with the System 360, the programming must be done completely and carefully—as the computer jargon cautions us "Garbage in—Garbage out."

Efforts to improve the quality of programming have involved changes in administrative organization, grading systems, course content, core requirements, general education requirements and teaching systems. Some of these changes have snowballed to the point they can be called trends. Some have helped, some have not. While the focus of this change must be directed toward (a) the student's need and (b) the most effective system for the faculty to meet this need, neither of these foci stresses efficiency. Let us examine some of the major trends in terms of student response and faculty effectiveness, keeping these prime objectives distinct from those of lesser value— number of student majors, course appeal or popularity.

The agricultural student of today is seldom a farmer, is no longer provincial in attitude and interests. He comes from a varied background and with varied preparation for college work. He must, in 4 years, become a good biologist, integrating physics, mathematics, chemistry and biology toward a comprehension of specific organisms and their interaction with natural and artificial environments. He may emphasize one of a number of career objectives—but in no sense is he less a student than his counterpart in other colleges or majors.

Structure

There is a restlessness among faculty and administrators that for some universities has resulted in major reorganization with a view to assembling in one unit the so-called pure sciences or pure biological sciences. In general, agriculturists are not behind this move for so much of their success has been achieved through coordinated basic-applied approach to problems. While the scientist's primary interest in this move is probably to promote research—improved grantsmanship—the liberal arts faculty and others teaching what they regard as untainted subject matter favor this type of organization from a purist's stand-point. In short, most agriculture colleges have found it difficult to muster enough votes to defeat such reorganizations. Our concern may be how to prevent such changes from occurring or, perhaps more commonly, how to modify and redirect our teaching program after a change is made. It is just possible that we can take advantage of such situations.

Reorganization raises issues that must be dealt with. One that im-mediately comes to mind is relevancy. We have faced the "relevancy gap" before with regard to physics and other basic courses, a dilemma that seemed to dissipate with development of "Ag. Physics" course, for example. But this course and others like it quickly proved medi-ocre and conveyed a poor impression of our college and our student majors to the whole academic community. We may face similar prob-lems in dealing with a college of biological sciences. Courses in bio-chemistry, for example, may relate less and less to the higher plants and animals so important in agriculture. And no matter what our peculiar needs may be, I doubt that the college of agriculture will ever be generally successful in coaxing science departments in other colleges to prepare and present courses especially for agriculture. They are not enthusiastic about such courses and neither am I. New approaches are needed. We must educate the student in the basic sciences, but his strength and the strength of agriculture itself still require that both student and teacher have their respective feet firmly on the ground.

What's in a Name?

For several years, there was a bandwagon move toward the tri-option system, consisting of business, technology, science. I suspect that it was designed primarily for efficiency and appeal, but some ascribed to it considerable academic value as well. Measured in terms of num-ber of students, I doubt that it had any long-term effect; more impor-tant, one might question if it really enhanced the education of the student or improved effectiveness of the faculty. The courses were the same—the old prerequisites were still there.

The same argument might apply to renaming colleges of agricul-ture. Names may have some effect in advertising the product; but, after the product is sampled . . . ? New names should not merely glamorize, they must reflect a fundamental change in personnel, courses, and direction.

General Education vs. Professional Courses

Our agriculture major is now faced with more exposure to sociology, philosophy, history, arts, language, government, mathematics, and economics than in earlier times. Few agriculture faculty dispute their value in helping a person adjust in this complex society, but increasing general requirements has reduced the number of professional courses that a student may take. In retrospect, this has been a blessing in convincing many departments that fewer courses can still adequately develop a student to the desired level of learning, that the self-centered curriculum is *not* the only way to assure a well-trained individual. The net result has been a trend toward streamlining the curriculum, separating the chaff from the grain. Currently, too, there is more willingness on the part of faculty to team up, to cross departmental lines, to present a good, solid multi-lateral course. This makes sense.

General Studies

There has been a tendency to include a general studies program in agriculture, usually directly administered by the dean of resident instruction. It was developed as a home for the undeclared major—the searching student—but has, I am sure, become a sort of sanctuary for avoiding tough courses or departmental requirements. Measures to assure quality and direction are just as important for general studies as for a departmentally oriented program.

New Programs

New areas of concentration are being created in colleges of agriculture, but in no uniform way. These, in particular, include honors programs and foreign study. The honors program is designed to spur on the superior mind and is a good idea, but as long as great emphasis is placed upon grades earned, students may be reluctant to enroll in an honors section. Perhaps the pass-fail system being adopted in some other course areas would be more appropriate here, if combined with an entrance requirement that would assure honor students in honors classes.

International programs raise some questions. Is the program designed to train foreign students to return to their homeland? Is it to train Americans for foreign service? These are two very different objectives, requiring different programs. It is obvious, too, that a student cannot specialize in more than one continental area: There simply is not enough time to study the language, politics, social systems, history, and agriculture of more than one. Curricula involving direct foreign experience for U.S. students, once the logistics problems are solved, would seem to be of special value.

Teaching Systems

Teaching machines are becoming sophisticated. First adopted in language labs, modifications are showing up in a few agricultural courses. The recent business mergers of such electronics firms as C.B.S. with publishing companies suggest that we can expect much more in the way of both equipment and technique. I suspect that there are many courses in our colleges that would be enhanced tremendously through use of such equipment. Financing the required equipment in the face of small class enrollments is difficult to justify on efficiency alone, but there is no problem justifying it on the basis of teacher effectiveness. Agriculture could lead in developing unique approaches to teaching because we have such a diverse array of subject matter and biological organisms from which to choose.

Science Core

It is generally recognized that certain courses are in the "must" category for most students in agriculture, but there is quite a gap between the recognition and action. Personnel problems, tradition, lack of facilities seem to oppose needed change, and these forces cannot be dismissed lightly. The action is painfully slow.

CONCLUSIONS

There are many changes that will grow out of efforts to solve the problems of the day. Textbooks, or the paucity of them, teaching machines, teacher evaluation all affect course quality; and the curriculum is no better than the quality of its courses.

Such things as pass-fail options, general education requirements, biology cores, accreditation standards, entrance requirements, advising systems, administrative structures all concern—occasionally plague—those involved in planning curricula. The levels of student

attainment in math, chemistry, and physics, before and during college, must be taken into account.

Experts predict serious food problems by the turn of the century—
no country excepted. The changes effected in our teaching between
now and the end of the 1970's will have prime impact on those individuals who reach the peak of professional competence in the years
1990–2000. Is not this a critical time?

It would be unusual, indeed, if changes imposed on colleges of
agriculture offered no opportunity for gain. Even seemingly dele-
terious change can be exploited by the imaginative. Those content
merely to react to change will be viewed as chronic complainers.
Those eager to seize upon change as an opportunity for improvement
will be the architect's of agriculture's new era.

DUANE ACKER

In 1969 there were approximately 100,000 students studying agriculture and related sciences in the nation's colleges and universities.
Though precise figures are not available, probably 60,000 of these
are in what could be termed hard core agriculture. These students
represent a significant responsibility. We recognize a division of responsibility in this task of educating students in the agricultural
sciences. The first is among the various institutions—junior colleges
vs. state colleges or private colleges vs. the land-grant universities. The
second is between the agriculture faculty of the institution and the
faculties in mathematics, English, physical sciences, biological sciences,
and other disciplines.

RECENT CURRICULUM CHANGES

The years 1946 to 1952 saw a big infusion of new, recently trained
PhD's into agricultural faculty. Their orientation for the most part
was biological. The importance of the basic sciences in helping to
understand biological phenomena was paramount in their minds as
they considered curricula. They were noticeably lacking business ex-

perience or business orientation; essentially none had significant experience in nonfarm businesses or had been successful in food production enterprises. One could say that these faculty were and are biological consultants. Their responsibilities have been to teach and do research in the biological disciplines, to cooperate with other biologists, to read the biological literature, and to study biological phenomena. Very few have served as consultants to feed mills, fertilizer plants, sales organizations, or farm production units. I believe this collective faculty characteristic had and continues to have a major influence in giving our undergraduate curricula a heavy orientation to biology and a paucity of business and management courses; e.g., mathematics to support biological and physical sciences rather than to support management technology.

From about 1956 to 1965 a high proportion of colleges of agriculture developed, within their majors or curricula in traditional agriculture, options in science, production (or technology), and business (or industry). This action simply recognized that 25 to 50 percent of the graduates in a given area—animal science, agronomy, or similar curricula—take their first employment, and many remain, in the nonfarm industries allied to agriculture. Faculty decided their graduates should be better equipped in such topics as business law, accounting, and personnel management.

Usually, to make room for each business course, a nonmajor agricultural course was omitted from the curriculum—the animal science students pursuing a business option took very little agronomy, the agronomist was able to get only the introductory course in animal science. This shift was hard for some faculty in colleges of agriculture to accept, but I believe most would agree that this movement was beneficial. We did observe real difficulty in going far enough into principles of business in the business options or in having enough enterprise management in the production options.

GENERAL AGRICULTURE CURRICULA

In the late 1950's and early 1960's there developed on a few campuses broad curricula in agriculture, such as agricultural science or agricultural business. These seemed to arise as a result of curriculum committee discussions or at the suggestion of administrators. The main reasons advanced were that students often do not know what discipline within agriculture they want to major in and that the curriculum in agricultural science should suit a young man planning on graduate work.

Very few of these "broader agriculture curricula" in major universities have succeeded except as a feeder to departmental curricula. I think the reason is that they generally do not provide a departmental home. Students in our colleges of agriculture, perhaps more than other students, apparently want to identify themselves with a particular discipline, a particular group of faculty, and a particular occupation or profession. We might call them job-oriented. Studies have shown that our students come to the university to equip themselves to specialize in a particular field, not just for the sake of "getting an education."

Some faculty have suggested that we may be moving toward a common undergraduate curriculum for all students in colleges of agriculture, permitting specialization only at the master's degree level. But we forget the student and what motivates him. In my opinion, this move is not likely to occur, for several significant reasons:

- The job-oriented motivation of our students.

- The major improvement that has been achieved in the level of education students have when they leave high school, permitting specialization and professionalism at the B.S. level.

- The unlikelihood that society will be willing to pay the cost of prolonging education; society would perfer to permit specialization early and then invest in continuing adult education. Changes in technology and talents required of people and rapidly developing programs in continuing adult education, designed to accommodate professional workers' need for new professional skills, support this premise.

INTEGRATION WITHIN THE UNIVERSITY

The next step, in my opinion, will be toward undergraduate "majors" and "minors," in place of traditional "curricula" in colleges of agriculture. College of agriculture faculty have repeatedly rejected curriculum accreditation, for many reasons. Therefore, fixed curricula patterns need not be maintained for that purpose. Increased flexibility in program planning is usually desired and justified by students and their advisers. The major and minor system, already in use at Iowa State, permits the student and his adviser flexibility, allows identification of a second or third area of specialization by the student, and may well encourage increased curriculum interaction between colleges of agriculture and departments in colleges of arts and sciences.

INTERNATIONAL AGRICULTURE

It has been estimated that 50 percent of our graduates will travel or work in a foreign country during their productive lifetime and that 30 percent will have specific employment in, or business relations with persons in, foreign countries.

Several colleges of agriculture have sponsored travel programs or programs of one-semester-at-a-foreign-institution for undergraduates; at the graduate level this is a bit more common. Undergraduate programs in international agriculture are spelled out in many agriculture college catalogs. The common experience has been that many students express interest during their freshman or sophomore year, but that not more than one or two in each class complete the full program.

Oregon State and several other universities report highly successful seminars, designed to expose agricultural students to international opportunities and challenges.

All this suggests that the technique for preparing American students for international agriculture is unlikely to be via separate curricula or even separate courses, but rather through (1) the development of an awareness through conferences and seminars, (2) the development of an international attitude by exposure to faculty who have served overseas, and (3) the development of international competence—and confidence in that competence—by teaching soils, or animal nutrition, or plant ecology on a world basis rather than on a North Carolina or Mississippi or Arkansas basis.

ACCOMMODATING TRANSFER STUDENTS

During the past decade most colleges of agriculture in states with a junior college system have worked diligently with the faculty and the administration of these junior colleges to mesh curricula offerings. Many have urged that junior colleges not teach technical agriculture because of its high cost, the difficulty in obtaining qualified teachers, and shortage of laboratory equipment. Where agriculture is offered at the junior college, the state university usually reserves the right in effect to certify the teacher and the course by giving or withholding credit in the professional curriculum when the student transfers. In some states the colleges of agriculture have offered the junior college video-taped courses, or at least conferences during the summer in which course outlines, workbooks, and methods of teaching might be discussed and shared with university faculty.

Several years ago Kansas State University spelled out the courses to be taken by those intending to transfer, a single list being set up for the whole college of agriculture. This required some compromise by departments. The list was taken to each junior college and adjusted to fit its course sequence so it could be printed in the catalog as a pre-Kansas State University agriculture course. It was felt the university might just as well decide what it will or will not accept *before* the student enters the junior college. These courses were guaranteed as acceptable for a three-year period.

In states where several colleges provide four-year programs, or at least the first two years of an agricultural curriculum, there have been some attempts to achieve uniformity among courses offered. The measure of success in this worthy venture varies.

COURSE CONTENT

Students in colleges and universities will increasingly demand germane and up-to-date course content. I think the present lack is one of the underlying causes of student unrest. Over the next few years I think we will see more rapid adjustments in course content initiated by the instructor, more department heads sitting in on classes taught in their departments, and more review of course outlines by department heads and departmental faculties.

To insure that course content be directed toward the mission of the department and the course, I think we will see a tendency for instructors to state the goals of their course rather specifically—10 to 15 specific pieces of knowledge, competencies, or concepts that the students should expect to achieve and that the instructor expects them to achieve. This will give the student and the instructor bench marks against which to measure progress. The instructor may also specify what he expects the student to bring into a course. In sequential courses, the concept a student is expected to bring to course 2 should mesh, of course, with the expected accomplishments of course 1. This device will result in many benefits to the instructors, but most importantly to the students.

I am amazed at the frequency with which agriculture faculty are not even acquainted with the person who teaches a prerequisite course in one of the basic sciences. We have a lot of work to do in this area. I believe we also lack adequate rapport with our high schools as far as curriculum is concerned. Two years ago we took our department heads to visit a high school biology department—it was most enlightening. I would propose that we encourage some of our faculty to spend their sabbatical leave teaching in some of the more progressive secondary schools in their respective states. They would probably teach biology, but experience in teaching chemistry, mathematics, physics, or vocational agriculture might also be useful. I recognize most will not meet formal state teacher certification requirements, but they could be identified as teacher's aides.

TEACHING EFFECTIVENESS

Probably 80 to 85 percent of the pressures that exist for curriculum change are "teacher oriented"—fifteen percent or less are subject oriented. In other words, the pressure for change results from the failure of a teacher to make his course effective and appropriate for the students who take it. Consequently the advisers of these students pressure for a "curriculum change." They will suggest, "We need a course with a little different emphasis," or "We need a course with less lab and more lecture." In most cases we try to be tactful by taking the committee route to achieve a curriculum change, when what we really need is a reassignment of teachers or reorientation of the teacher to the needs of the students he has in class. I think we will see, increasingly, that any new teacher of a course is thoroughly briefed by the department head on the students he will have in the class, the curriculum or curricula from which they come, the missions of those curricula, and the other courses in the sequence where his course fits the backgrounds and aptitudes of the students. We do a pretty poor job on most campuses of orienting new faculty to their teaching responsibilities. This handicaps their chance for success. Indeed, it is probably a matter of luck that we have as many successes as we have.

We also recognize the improved teaching effectiveness that results from identifying and rewarding effective teachers. At the Gamma Sigma Delta banquet at Texas A&M last spring the agriculture and veterinary medicine students gave a certificate of award to a professor of English for his effective instruction, an event that may improve teaching effectiveness in the total English department in the years ahead.

INDEPENDENCE IN CURRICULUM THOUGHT

In the decade of the 50's and for a few years beyond, faculty in some colleges of agriculture developed considerable self-consciousness about their colleges, their curricula requirements, the quality of their students, even their own disciplines. Perhaps it was because the 50's was accepted as the decade of the physical sciences. Perhaps it was due to the enrollment decline in agriculture after 1955. Perhaps it was Sputnik and the recoil away from "vocational curricula."

Some reacted by saying, "We are not concerned about numbers; we want quality," and implemented a very rigorous standard freshman year. Some said, "We are just as professional as the engineers," and added the engineering math sequence to all curricula. Some said, "We are not 'vocational,'" added 20 credits of social sciences and humanities, decreasing the number of courses in the major agriculture field.

In the 60's faculty attitudes changed. Effects of curriculum changes of the 50's were carefully reviewed and assessed. Society became concerned about "feeding the hungry world" and thus about agriculture. High investments in agricultural enterprises—farm and nonfarm—demanded competent agricultural professionals. Enrollments in colleges of agriculture surged—growth has in many cases been much faster in agriculture than in other disciplines. Faculty became more secure of their position on the campus and adopted new attitudes regarding curricula.

One college interrupted the collegewide requirement of engineering math and adopted a course in finite math for agriculture business majors and those in business options. Several colleges added undergraduate statistics as a requirement. A director has proposed to his faculty that a "level of competence" in basis communication be required rather than "six credits of freshman English." Several colleges of agriculture have added biochemistry to their plant and animal curricula; a few are pressing their chemistry departments to move biochemistry to a position earlier in their course sequences.

Agriculture faculty now are less concerned about how faculty in other colleges view them and their programs and students. They are designing curricula for *their students* and the sciences and professions *they will enter.*

CLASSIFYING AND EVALUATING PROGRAMS

There is reason to believe that colleges of agriculture will develop in the years ahead a system of classifying and evaluating academic programs. Traditionally, college of agriculture faculties have rejected accreditation in any form because of its stifling effect on curriculum change, the way it prevents experimentation in teaching and in curriculum, and because of the abuses that often result.

There are two major reasons, however, why classifying and evaluating programs will probably be seriously considered.

The first is to insure that students and their parents know what a curriculum is and what it is supposed to do. With agriculture available in many kinds of colleges and universities, with agriculture faculty varying greatly in size and training, and with the tremendous variation in the goals at various institutions, students may be badly confused as to what they will receive. Some curricula stress liberal arts, with a touch of agriculture, taught by faculty members with the equivalent of a master's degree or less. Others are pregraduate curricula, highly specialized. Still others are professional degree programs aimed at employment within a specific agricultural profession after a B.S. degree.

It is wrong to permit a young man to enter a pregraduate curriculum when he clearly wants and expects to obtain professional training for employment after two or four years. It is equally wrong to permit a young man who is seeking professional-level education in soils at the B.S. or M.S. degree level to enter an institution with only three or four agricultural faculty with little or no specialization in soils.

The second reason for classification and evaluation is to provide the college of agriculture with the leverage, now enjoyed by many schools of education, pharmacy, or nursing, for obtaining faculty, buildings, and operating funds. We are all aware of instances in which accreditation reports have forced administration, regents, or legislatures to increase funds for the accredited unit, under the threat of losing accreditation. And while these increased funds may be fully justified, agriculture often suffers by comparison.

2

Trends in Renewable
Natural Resources Curricula

JAMES S. BETHEL

Any meaningful discussion of trends in renewable natural resource education must be based, it seems to me, on the history of renewable resources use and education. At the outset, this country assumed the custodianship of a vast territory richly endowed. Much of the land was productive, in most areas water was abundant, fish and game were present in large quantities, and forests or grass covered large areas. These natural resources did not arise as a result of human manipulation but as a consequence of the development of natural ecosystems. At the outset, they were exploited without regard for a continuity of supply; indeed, they were sometimes considered obstacles to more effective use of the land.

Public sentiment for treating natural resources as renewable national assets did not emerge strongly until the last half of the 19th century, when professional forestry education also appeared on the American scene. The first two forestry schools in the United States were established in 1898, one at Cornell University and the second, the Biltmore School in North Carolina. In 1900 a forestry school was

established at Yale University, and during the first ten years of the
20th century sixteen additional forestry schools came into being.
Among early forestry schools were private universities such as Yale
and Harvard, private colleges such as Bates and Colorado College,
state universities such as the University of Michigan and the University of Washington, and a number of landgrant institutions. Instruction at these institutions was sometimes exclusively at the graduate
level, sometimes both undergraduate and graduate.

The early graduates of these institutions were concerned primarily
with the conservation and use of the native forests—protection
against fire, insects and disease, inventory of the resource, and the
development of means to convert the forest to goods and services.
The strictly custodial aspect of natural resource management is now
largely in the past so far as the United States is concerned. The resource manager today is more concerned with maintaining the continuity of the biological resource and optimizing its output in terms
of quantity and quality of materials and services. He is a manipulator
of biological and physical systems in which the outputs are products
for commerce and industry and social services for the general population. The constraints on the system fall into the domain of the
physical, biological and social sciences.

This, then, is the setting for renewable resource education. As the
objectives change so must the education itself change; indeed it is
now doing so at a very rapid pace.

Trends in renewable natural resource education reflect changes in
the institutional environment. Many universities give increasing attention to graduate education and to undergraduate preparation for
graduate education, which is in turn reflected in the synthesis of curricula. Two-year community colleges and junior colleges are beginning
to train technicians and to provide early preparation of professionals.
In the future much of the routine technical work that has been performed by junior professionals will be performed by technicians. Professionals will thus be relieved of the necessity of acquiring many of
the practical skills that were essential in the past. This makes it possible for the university to emphasize the science and mathematics
base of professional education.

Curricula in renewable natural resources are being revised to reflect
changing professional needs. The modern professional must be prepared to bring his task the power of the natural sciences, social sciences and mathematics. Incoming freshmen are better prepared at the
high school level than they used to be. Increasingly, schools and col-

leges assume that entering students have been exposed to modern high school physics, biology, chemistry and mathematics. Where this is not so, deficiencies are commonly made up in remedial courses. The problem for the faculty, not always an easy one, is to devise appropriate building blocks for erecting an education structure on this stronger base, but the best building blocks may not be so organized that they can be neatly assembled into a four-year package.

Biology is perhaps the most important of the sciences that are basic to resource education. Thus far, the major efforts to improve basic biology courses at the college-university level have been in developing courses around the unifying contribution of molecular and cell biology, courses that are then commonly deferred until the sophomore year to permit prior college level preparation in chemistry and mathematics. Unquestionably, these are usually stronger courses than the more descriptive botany and zoology courses typically included in natural resource curricula in the past. But it is not at all clear that we have as yet achieved the optimum. It is still common to find that the choices available from biology departments are either the modern biology with molecular and cell biology as a unifying element, or classical biology that is largely descriptive in character.

A recent analysis of several new biology core curricula* indicates that of the total time available in the introductory courses the fraction devoted to cell biology was about five times that devoted to ecology. Many resources biologists are beginning to ask whether it would not be appropriate to explore a second route into introductory college biology, a route that would utilize the concepts of ecosystem analysis as a unifying element. Such a course ought to be as rigorous and as quantitative as the one based on the cell. It might, indeed, require comparable preparation in mathematics and the physical sciences.

As professional preparation in natural resources is increasingly moved to the junior and senior years, there is a real need to consider the motivational properties of modern introductory courses. The increased levels of preparation in basic sciences tends to minimize a student's early contact with his professional field. What may be lost thereby is a persuasive basis for showing the relevance of the basic disciplines to long-term professional objectives. New mechanisms are needed that can underscore this relevance and motivate the resource-

*"Content of Core Curricula in Biology," Publication No. 18. Commission on Undergraduate Education in the Biological Sciences; June, 1967. American Institute of Biological Sciences, Washington, D.C.

oriented student to achieve a considerable competence in the basic disciplines as a necessary prerequisite to sound professional preparation. A premedical student may see himself as a manipulator of cells, tissues, and organisms and remain excited about biology and medicine while acquiring the necessary background. A prospective resource scientist, on the other hand, is likely to think of himself as a manipulator of ecosystems and will find biology more relevant to his ultimate career goal if he is introduced to it within the framework of the ecosystem. It has sometimes been argued that a systems approach to ecology is too difficult for an introductory course and ought to be delayed. But it is difficult to see why Odum's biomes, Holdridge's life zones, population dynamics models or energy flow models are any more difficult to understand than protein synthesizing enzyme systems or the Watson-Crick model.

Similar problems face the faculties as they select curriculum components in mathematics, chemistry and physics. Most programs in the resource fields today require a substantially higher level of mathematics preparation than was the case five or ten years ago, yet all too frequently the best mathematics sequences, taught by the best faculty members, are those devised to meet the needs of mathematics or physical science majors. Here, as in biology, the faculty must choose between poorly organized and inadequately taught service courses and the better modern courses that lack relevance to natural resources. Preliminary recommendations have been made by a committee of the Commission on Education in Agriculture and Natural Resources, but the courses described are offered by far too few mathematics departments. A similar situation exists in chemistry and physics.

The trend in natural resource curricula is to reduce the number of courses and to improve the content, taking advantage of the higher levels of preparation in basic science. As still more appropriate basic science courses are developed and available, this trend may be expected to accelerate. A very real limitation on improvement is faculty obsolescence, a problem that is, of course, not unique to the natural resource field. To date there appears to have been much less effort expended in trying to solve this problem in agriculture and of natural resources than in many other fields of science and technology. It should be aggressively attacked.

One cannot conclude without some reference to the increasing efforts of natural resources faculties to contribute to general education in the universities. The provincialism of colleges and schools of forestry and natural resources is rapidly breaking down. Students from other parts of the campus are getting interested in the study of natural resources as one of the liberal arts, a demand being met through a growing list of courses for nonmajors. It is important that this trend be continued and that universities increasingly recognize that resources faculties gain strength from the university environment of which they are a part and, at the same time, feed back into that environment their own special interests and competencies.

JOHN F. HOSNER

We think of natural resources in terms of the currently recognized programs—forestry, wildlife, fisheries, range management, watershed management, soil and water conservation, and recreation and park management. Except as otherwise noted, my remarks will deal with aspects common to essentially all groups.

Our population is increasing rapidly and is becoming increasingly urbanized. We will soon be a nation of urban dwellers. Both these trends exert important influences on the use of renewable natural resources. We are experiencing sharply accelerated and competing demands upon them. For example, demand for recreation, in large measure a by-product of urbanization, is increasing at approximately 10 percent per year with no evidence of slackening. All this makes it mandatory that we emphasize new educational dimensions in the renewable natural resource fields.

It is imperative that the scientific foundations and equipment for decision-making be incorporated into the resource manager's bag of specialized training. Linear and dynamic programming, simulation, and other complex approaches made possible by computers must become a regular part of the well-educated resource manager's education. We must, at the same time, update and modernize the basic science courses.

We observe increasing competition for land uses and staggering developments in science and technology. We must be prepared to apply this new science and technology to our renewable natural resource problems. If resource managers do not take advantage of these opportunities, then engineers, economists, and other outside specialists will take over management of our renewable resources, while the trained resource manager becomes relegated to the technician role.

While we provide training in specialized, often sophisticated, techniques and education in the basic physical and biological sciences, we also must provide a broader education for the resource manager of the future—an education that will permit him, in the words of Henry Vaux (1961), "to perform the strategic function of integrating, interpreting, and linking forestry to the rest of society in a fashion most meaningful for that society." Dr. McArdle (1961), former Chief of the U.S. Forest Service, succinctly stated the problem when he said, "Now, as in the past, I firmly believe that most such policy decisions [those dealing with the use of forest land] are made not by foresters, but by legislators, executives, financiers, engineers, and men of other disciplines and orientation. If ever there is a challenge to foresters, it is to escape from narrow technocracy and to engage actively in the practice of political science and business management."

We have hardly begun to analyze, much less deal with, the education of natural resources specialists to the broader role of public leadership and national policy development. There is a distinct lack of breadth in the training of wildlifers, foresters, recreationists, and the like. The emphasis seems to be on technical subject matter with but limited requirements in humanities and social studies. For example, only one third of the schools listed in the Dana and Johnson report (1963) on forestry education in America required any work in the humanities and only one half required social studies.

The need is clearly evident for truly educated potential leaders in the renewable natural resource areas—individuals who can assume a larger leadership role in the broadest sense.

We must, of course, be cognizant of today's immediate needs. Foresters, for example, should be able to inventory a tract of timber being offered for sale. Wildlifers should be familiar with the common techniques of the trade, for example, animal aging techniques—the list could go on and on. I think it is obvious, however, that emphasis must be shifted from vocational training to professional education, and from instruction in current practices to teaching basic principles.

It has often been asserted, that the half-life of a Ph.D. today in the scientific discipline is 5 years. I do not know what the half-life of our graduating resources managers and scientists is in today's environment, but I submit that it is correspondingly short.

Increasingly, we see schools substituting mathematics, economics, English, political science, and similar foundation subjects for the technique courses. As these changes occur, educators increasingly encounter reactions from employers, what they say, depending on the agency they represent and to some extent on the position they occupy. For example, if one were to generalize about forestry, it would be safe to say that the top administrator feels emphasis should be shifted to fundamental concepts, while the area forester is more interested in hiring a graduate proficient with field instruments and techniques—the latter sees little need for calculus, political science, and similar courses. Consequently, we in forestry education often find ourselves heeding administrative personnel only to get static from the graduate's immediate supervisor because we have turned out a forester somewhat less than proficient in handling the techniques of the trade.

The situation might be summarized as follows:

• The environment for the management of the renewable natural resources is changing. There is an increasing need to recognize and integrate in the most efficient manner all competing demands for the available resources.

• This increased competition will demand increasing skills in modern management decision-making that require sophisticated application of computer techniques.

• The increasing fund of basic information to be dealt with will demand that we pare to a minimum the amount of our curricula devoted to vocational skills. Skills that can be as effectively learned on the job must be postponed to make room for the more fundamental courses.

• With our increasing "people" problems there is a distinct need for a broader-based training in the humanities and social studies—studies that will provide the background for natural resource managers to integrate, interpret and link their interests to the rest of society.

REFERENCES

Dana, S. T., and E. W. Johnson. 1963. Forestry education in America, today and tomorrow. Society of American Foresters, Washington, D.C. 402 p.

McArdle, R. E. 1961. Challenges in forest land use, p. 47–54. Proc. Fiftieth Anniversary Celebration. State University, College of Forestry at Syracuse University. April 12–14, 1961.

Vaux, H. J. 1961. Challenges in forestry education, Proc. Fiftieth Anniversary Celebration. State University, College of Forestry at Syracuse University. April 12–14, 1961. p. 19–29.

> DeWITT NELSON

If one is to give direction to the future, one must first examine the present. Population, resource demands and environmental problems, present and future, are inseparable. To build for the future we must create an alert and knowledgeable society willing to bring reasonable balance among exploitation, rehabilitation, and preservation of the nation's natural resources. This will require a high degree of inter-action and understanding between our educational, social, economic and political institutions.

Technically and scientifically we can solve most of our problems. But can we as a people, and as a nation, suppress our personal and in-stitutional ambitions to the point where we will be willing and able to work together for the common good?

Can we educate enough of our people to understand the delicate balance that exists between man and his environment? What are some of the problems confronting the educational institutions? What are some approaches that may be initiated to improve public awareness and understanding so that our social and political institutions will act more promptly and responsibly?

POPULATION

As we look about us and examine the trends of population growth with its ever accelerating demands for more goods, services and the amenities for living, we cannot but wonder when will we reach satura-tion. How long can our planet meet the demands placed upon it? Our natural resources are finite—without control the population is infin-ite. Consequently we find ourselves approaching a serious imbalance between man and the resources upon which he is dependent—the soil, water, air, plants, animals and minerals. It is obvious that we are on a collision course between the reproductive capacity of man and the pro-ductive capacity of nature.

The best information available indicates there were about 250 mil-lion people on Earth at the beginning of the Christian era. It took 1600 years for the population to double. There are six times that number today—less than 400 years later. It is predicted that the 3 billion on earth now will increase to 6 billion by the year 2000.

A child is born about every 9 seconds here in the United States; every morning there are more than 9,500 new mouths to feed. To do this will place ever-increasing stress on our nation's natural resources. More than 90 percent of these people will have little or no contact with or relationship to actual farming, harvesting of forests or mining of minerals. They will not be exposed to many of the biological relationships between man and the resources. Most will live in urban areas.

DEMAND

In the next 30 to 35 years this nation must double its food production on our limited land base. Besides food and fiber each person needs for a lifetime of modern living more forest and mineral products, more and more parks, playgrounds and other recreation areas, increased space for travel vehicles, homes, schools, commercial establishments, and the like—all requiring land surface. Already less than 50 percent of our total land area is used for farms and ranches. Good agricultural land must be protected from encroachment by other demands that could be met on less productive land. To solve such problems will require the support of a public generally knowledgeable about agriculture, natural resources and their development and use.

RESOURCES

Our land base is fixed. Our once overwhelming abundance of natural resources no longer exists. Our consumption outpaces our productive capacities in many areas. The increasing population confronts mankind with frighteningly difficult and complex problems. The competition for land and water between agricultural and nonagricultural uses is keen. The managers of resources have too often ignored the long-term effects of their actions and have inadequately recognized the interlocking nature of the situations with which they deal.

The traditional concepts of natural resources are changing. People are placing new values on resources for both their products and services.

Man is not dominant over Nature, yet he continues to ignore many of her warnings. He continues to dump an ever increasing array of garbage and pollutants on the land and into the air and water with little concern for the ultimate threat to his welfare or survival. He continues to build water impoundments without first controlling the erosion on the upper watersheds. Over 2,000 such reservoirs in the United States are now useless impoundments of silt, sand and gravel.

Raw sewage and industrial wastes continue to hasten the eutrophication of streams and lakes. We reduce the assimilative capacity of the receiving waters, thus destroying acquatic life and degrading the water for other needed uses, including the esthetic values.

MAN'S RELATIONSHIP TO HIS ENVIRONMENT

Nature has the capacity to produce if we work with her. But if we are insensitive to her demands, we only create new dust bowls, eroded hillsides, sewer-like rivers, and cesspool-like lakes. We can destroy the food-chains for essential living organisms, overload the atmosphere with hydrocarbons and other toxic elements, and ultimately destroy the essentials of life.

It is man's choice. Man is governed by the same laws that govern all living things. He must learn to be sensitive to Nature's warnings— "the handwriting on the wall." He must learn, and learn quickly, to work with Nature if he is to continue to enjoy her abundance.

These choices are within man's power. He must harness his technology. He must analyze the ultimate consequences of his technical and social actions. We can better predict the long-range results, good and bad, of technological programs and actions than we can the actions and reactions of people. True conservation will be recognized as sound national policy when the long-term benefits and detriments are adequately weighed against the short-term profits and costs, and when the public interest is adequately considered with the private interests.

We must establish priorities of expenditures to make and keep the world fit for living. It is easier to get money for damming rivers than managing watersheds. It is easier to harness the power of technology for a moon landing than it is to coordinate hundreds of industries and units of government and millions of people to control waste and to use well the land. We can better solve our biological problems if and when we can solve our social problems. Our environmental problems require a combination of social, economic, political and technical solutions. We must marshall the strength of people in our pluralistic form of government behind programs to support environmental protection and control. It has been said that "What we *ought* to do is now more important than what we *can* do. In our tradition the *can* question must be answered by an informed and participative electorate."*

The development of an informed and participating electorate extends beyond the province of the scientist and technician, particularly in this age when the majority of our people live in high density areas. If we are to have a sensitive and informed public, we must teach an awareness and understanding of the environment. We must place educational emphasis on the biological and sociological disciplines for all students regardless of their specialization.

Public attitudes and public financial support carries weight at the decision-making level as to how both public and private lands and resources are managed. Governments are responsive to public opinion. Industries are becoming more sensitive and responsive to public demands. Therefore it is important that public opinion be based on factual information and intelligent interpretation by a knowledgeable electorate.

The recent power shift has placed the resource decisions in urban hands. By integrating the principles of conservation and utilization in their broadest sense throughout the various curricula, the major resources issues can be more properly understood and evaluated. This should lead to better policy decisions in both the public and private sectors.

CONFLICTS AND CONTROVERSIES

There are many controversies within the broad field of conservation. Resolving these conflicts becomes more difficult as the problems become more complex and as individuals become more specialized. A well informed lay person may be more appreciative of these problems than a specialist in a single resource area. I fear that this will become more and more the case as our population becomes more divorced from the land. I also fear it as we train our students in more narrowly oriented areas of specialization.

*Dr. Chalmers Roy, Dean, College of Science and Humanities, Iowa State University.

Recently a student came to my office with a real concern about the lack of understanding and appreciation of environmental problems by students—in engineering, economics and industrial administration—with whom he associated. From his comments I could arrive at only one conclusion—that in their course work the students were receiving no exposure to the biological sciences, to the facts of the world about them and to their dependence upon it. Yet those students will help formulate the public policies in the future. A second student, who grew up near one of our popular lake resort areas, made the statement, "I knew something was happening to our area but I didn't know what it was. Since taking your course I now begin to understand the inter-relationships and interactions that are taking place." What am I teaching? In the catalog it is titled Forestry 160. I can better describe it as "Resource Problems, Programs and Issues with Emphasis on Inter-disciplinary Relationships."

The failure to provide students with an understanding and aware-ness of man's impact on the environment can only result in making more difficult a coordinated action toward solving our environment problems. Man's failure to fully evaluate the consequences of many of his scientific and technological developments will create new and more difficult problems. Piecemeal attacks to answer specific questions may be essential, but they fail to solve the total problem. They also fail to recognize the interlocking nature between many piecemeal solutions. For example, the careless use of pesticides, removal of fence rows and clean cultivation have had serious impact upon the pollinating insects so essential to many agricultural crops.

FUTURE NEEDS

Until we are willing and able to manage our resources as a community of resources, rather than independently as individual resources we will be unable to achieve the environment we strive for. Too often we find the managers and disciplinarians of one resource competing with each other, or even worse, ignoring each other. I have spent most of my life in resource administration dealing with disciplines and their interactions as they effect both public and private points of view. As a consequence I have a strong conviction of the necessity to provide students with a firm interdisciplinary foundation.

I vigorously support the quest for specialists and research in every discipline. But in each we also need broad-gaged generalists who can effectively communicate and coordinate the interlocking relationships. Both the specialists and the generalists are becoming increasingly important components of the total resource team. Their orientation must reach beyond the resource of their interest. They must be people-oriented. They must be able to evaluate the social needs and consequences of their every action. We need to develop comprehensive courses, seminars and other educational devices tailored to provide an essential understanding of the interrelationships among all competing components of the resource complex. We must provide the necessary insight to deal convincingly with potential and actual conflicts in resource use.

EDUCATION

The nature of environmental improvement programs is interdisciplinary. Therefore materials exposing the student to his relationship to his environment must be incorporated into a wide array of courses—physics, chemistry, biology, geology, meteorology, forestry, agriculture, humanities, economics and the social sciences.

Courses in the physical sciences can well apply their principles to environmental problems. The concept of man as a part of a biological ecosystem should certainly be stressed in all biology courses. The concept of man's capability of operating with, rather than subduing, the dynamic forces of nature should be routinely included in agricultural and allied technological courses. The record of man's experiences in solving, or failing to solve, problems of adaptation to his environment could be included in history courses. The dynamics of conflicting purposes of resource utilization could be incorporated in economics and political science courses. The development of systems governing individual and group behavior with respect to natural resources is appropriate in the social and behavioral sciences.

Fortunately, there are many alternative routes through which man's relationship to his environment can be channeled, some of which are less costly than others, even though the end realized is not greatly different. In fact, environmental improvement must primarily be achieved by modification of activities undertaken for economic purposes. Therefore, the weighing of alternative values and costs are at the heart of the problem of environmental improvement.

L. C. WALKER

To the seventeenth century Englishman, the steward was the manager of the estate. He was a conservationist in the Pinchot image, the wise user of the resources. Thus, the steward was, and his counterpart today is, concerned with production of raw material for people to use. That concern must also involve the inerrelationship between those resources and those people. In a word, that interrelationship is economics, the *iconaea* of the Greek and the stewardship of the Anglo-Saxon.

JACK-OF-ALL-TRADES

Only during the past half decade have I been much concerned about curricular matters in forestry education. But in this period, my position has changed—I like to think it has matured—from a feeling that every professional forester must be poured from the same mold to the conviction that diversity in curricula is essential to solving the problems ahead. Twenty years ago on the 180,000-acre Sabine National Forest there were but two of us professionals. It was fortunate for the Forest Service that we knew some physics, some economics, some report writing, some law, some geography, and a whole lot of silviculture. Neither of us knew much history, accounting, sociology, game management, recreation management, range science, or engineering. Perhaps we really did not need to.

Now, the staff of that same National Forest, but with 30,000 of its acres under water, includes nine professionals. Were they all cut to the same pattern as were we two earlier foresters, little would be done about recreation management on and around that fine new fishing water, the unmanaged cattle would still be in trespass, wildlife would yet be poached, and the ledger books would be a garbled mess. All nine of these people who now manage this forest are and should be foresters, yet among them must be the skills to manage all of the various resources of the multiple-use concept.

Obviously, there is not room in a forestry curriculum to equip all graduates as specialists in every branch of our profession. Hence the core curriculum, which lists the essential courses that every forester should have as basic background, has been introduced. At some schools, it's as little as 15 semester credit-hours of forestry. At ours, it is 43 hours of forestry and another 53 hours of humanities and pure science. Thus some students elect additional studies in forest management, forest recreation management, or forest game management. Others may prefer an individually tailored program that permits minors in economics, business, journalism, computer science, physical or biological sciences, or general conservation. I would not call graduates in these options specialists or experts—in time they may become that; rather, they are foresters with additional work in certain fields. More than jacks-of-all-trades, but not quite masters of one!

Faculty response is a recognized hazard when the curricular mold, albeit not made of fragile clay, is broken. What professor does not think his course is *the* course, whose omission, as some wag has said, would be like sending the graduate naked into the world.

SILVIBUSINESS

Agribusiness was a brand new word in agricultural schools just a few years back. With its birth, both as a term and as a livelihood, the agricultural curriculum, along with the students in the schools, entered a period of greater sophistication. No longer need the agronomist be an educated farm boy—to sell fertilizer, to manage dairies, and to manufacture cheese requires men with business ability. Educational emphasis, then, moved from farm to marketplace. If the agricultural college was to stay in business, it needed to provide education to accomodate business, lest the schools of business do so.

Forestry educators recognized the wisdom of the agricultural educators' movement into business. A couple of schools developed curricula that appear very like branches of a collegiate business department. The danger, I think, is that too many students may be enticed into the business of forestry, arguing that forests can be managed from offices and by computers, at the expense of "on-the-ground" management. Others believe the emphasis on the entrepreneurial aspects of forestry—or silvibusiness—may lead to exploitation that takes into account neither the responsibility for maintaining the resources nor for serving the public.

INTO THE WOODS

Sidney Lanier, poet laureate of Georgia, wrote, in addition to the "Song of the Chattahoochee," which every Georgia schoolboy learns, an Easter-time verse, more lately set to music: "Into the Woods My Master Went."

Our students are not going into Judea's groves of olive trees, but they must be prepared to work in the forests and to feel comfortable there. If there is any one thing I would want of a young forester, it is an ability to observe and then to diagnose the condition of the forests and, having done this, to prescribe for their management. The crucial issue may be managerial, or it may be nutritional, pathological, or entomological. The curricular material is obvious. The program must be well-rounded, requiring basic and applied courses in these fields.

We believe a forest-oriented program begins with the freshman in the forest, not in routine subprofessional laboratory exercises but in direct exposure to management problems, mensurational data collection, or studies of mycorrhizae on roots of seedlings.

Silviculture is not an exercise in timber marking, but applied ecology in a dozen timber types—for our men both in the South and in the Rockies. The forest serves as the forestry student's laboratory, the timber type his practice cases, while the increment borer and the spade are his stethoscope and thermometer.

THE CRITICS

Forestry curricula get, in my opinion, more than their fair share of criticism. This comes, I note, from (a) educators, who are talking about their own schools; (b) foresters, reminiscing of their own college days "way back when"; (c) employers stung by a couple of unpromising appointments; and (d) leaders in the profession who have not been on a campus in a decade.

Like most school faculties, we are sensitive to criticism and anxious to mend our ways if necessary. Because we wondered if ours was one of those schools—they are never named—not providing for the needs of industry, we borrowed from one of our state's largest wood-using enterprises their Manager of Forests. This man, recognized across the South as a leader of men and a leader in resource management—and a critic of forestry education—sought out every nook and cranny of our program to learn what we were, and were not, teaching. To his surprise, and ours, only one subject was lacking: wood procurement! Our consultant then graciously outlined in detail a senior-graduate level course for which we are now endeavoring to locate a qualified teacher. It will join our course in forest law.

To the reminiscing foresters, just a word: course titles may not change, but material covered does. Hence, radioisotopes are tools in soils and silvics, data processing centers are labs for mensuration, policy enters the curriculum at the freshman introduction course, and economic geography replaces a lot of tedious memorizing of vegetative cover types in regional silviculture.

To the educators, my sympathy: It is not easy to persuade established faculties—long removed from the practice of forestry—that changes are in order. It is equally difficult to dissuade the young Ph.D. who is convinced he knows precisely the proper curricular composition.

TECHNICIANS AND TECHNOLOGISTS

Short programs—one to two years—in natural resource management are designed to develop technicians, while four-year-plus programs educate technologists. The former are trained for the "how to do it," the latter learn also "why you do it." Both are concerned with techniques that, due to the knowledge explosion, are proliferating logarithmically. More extensive curricula are inevitable; salaries for technological specialists will make a five-year program worthwhile. I predict that for foresters, in the not too distant future, the lowest professional degree will be the M.F.

The program recommended by the Commission on Education in Agriculture and Natural Resources for managers and scientists in renewable natural resources errs, I think, in having too little technological preparation. Perhaps it assumes the material to be subprofessional technician training and, therefore, not professional or just not important. Only 21 credits are allocated—and I assume that includes pathology, entomology, and soils. Can a man with so little technological eduation handle technological problems? If the man in this new age is to manage nature for society, he must participate in prescription as well as description, and that takes intensive technological preparation.

In contrast to the technician who works with "things," the technologist will work with people as well as things. While we worry about the forester needing humanities courses in order to "better get along with people," it is, rather, to lead and influence people that he is called. For professional resource management decisions must be based more upon technical knowledge—the objective—than upon people preferences—the subjective. The opinions of special interest

groups—whether Sierra Clubers or lumbermen—are important subject-
ively, but woe unto us if foresters make too many of their judgment
decisions on that basis. Humanities are important, but they are more
important for understanding people than for getting along with them.
The forester as resource manager needs this understanding, too. Thus
he operates at the "interface" of science and sociology.

RESOURCES

Man is unique in his ability to alter the environment. It is his privilege
to reshape it to his needs. Hopefully, any such reshaping will be clean
and orderly, because the land:man ratio is rapidly falling. If we labor
well, we are masters of the ecosystem upon which we depend; if not,
we are its slaves. Either way, we are a part of a changing environment,
with consequent alterations in the earth's resources on land and under
the land, on water and in the water, and air. Some form of those resources is universal, in walled city and unfenced wilderness.

I am not emotionally impressed with propaganda photographs in
publications of the Government Printing Office, the Ford Foundation,
and *Life* that depict the appalling conditions of city slums and, by
implication, blame the captive viewer for the situation. Rather, I am
depressed that people—kinsmen, if you will—should be so unconcerned
or lazy—in a society that possesses so much time and know-how, as to
allow such conditions to occur and continue.

Because individuals living in social misery appear unwilling or unable to improve their environment, professionals will no doubt be enlisted. Many of these professionals may be educated in the management of renewable natural resources, and their responsibilities will
likely include man, men as a community, and the environment of that
community. How to utilize the waste of the community brought on
by the economics of incomplete consumption, the waste of the fac-
tory and the waste of man himself, may be in the next few years the
crucial assignment of natural resources managers.

Foresters are a breed of planner, the resource engineer who under-
stands the biology of ecology, the human nature of ecology, the economics of ecology, and the manipulation of ecology. Conceivably the
manager will conclude that air pollution will eliminate plants through
reduced solar radiation long before it eliminates man, or that stream
and air dilution is inevitably pollution. Because production of goods
is not going to be curtailed—for environmental preservationists, like

the rest of us, must use the goods that begin as raw materials—especially careful management decisions must be made. The professional renewable resource manager will develop ecosystem capabilities and prescribe the tolerance levels of change for the components of the environment.

More parks, lakes, residential areas, campgrounds, rights-of-way, factories, and farm lands are rapidly reducing land available for watershed management, timber production, and game production and consumption. With less land on which to grow more wood for more uses for more people, management intensity must increase. The alternatives are substitutes for wood and substitutes for land. More effective management is the substitute for land.

THE FUTURE

Never has the future been more uncertain and, for the resource manager, brighter with opportunity. The Conservation Bill of Rights introduced into the House of Representatives calls for the "right of people to clean air, pure water . . ." It is the very uncertainty about our resources that makes the opportunities bright. Decisions must be made, policies established, and protection provided. To do this may call for the same kind of evangelism that motivated Gifford Pinchot and other early foresters.

Today's faculties of forestry, mostly educated in the 1940's, were in college because of altruism. Timber famine was yet the cry, and many a city boy left his "ghetto dwelling"—though we did not call it that then and the front walk was scrubbed daily before school—to remedy the situation. Students in the fifties and sixties, in contrast, were the first generation of American foresters who could concentrate on managing forests for profit. For the seventies, foresters must again be motivated by an altruism that, in managing both commercial and noncommercial lands, recognizes paramount public interests.

If the key phrase in the thirties was timber famine and in the fifties multiple use, perhaps in the seventies it will be *iconaea*—for foresters the bonding of man and land in resource management.

3

Trends in Biology Curricula

JOHN D. LATTIN

On October 14, 1965, three French scientists, François Jacob, André Lwoff and Jacques Monod, received the Nobel prize in medicine for their work on gene regulation carried out at the Pasteur Institute in Paris. One month later, Victor McElheny (1965) summarized the impact of this event on French science and in so doing quoted some critical remarks by Monod regarding the state of science in France. These remarks created quite a furor. As it happened, I was on sabbatical leave, working in the Laboratory of Entomology of the Agricultural University in Wageningen, The Netherlands, and heard some of this discussion.

The criticism itself was directed at the rigid structure of the French universities and how they interfered with interdisciplinary research. Monod himself had been discouraged from remaining at the Sorbonne, after completing his graduate studies because his work spanned two disciplines. Only the Pasteur Institute provided the proper environment for him to continue his research. John Walsh (1968) extended this criticism to include many of the European universities, at least with regard to molecular biology, and cited the University of Geneva as a distinct exception.

One of Monod's statements is pertinent to the topic here discussed, when he said, "in science, self-satisfaction is death. Personal self-satisfaction is the death of research. A man of science who is content with what he is doing and finds that all is going well—that's a sterile man. Unquietness, anxiety, dissatisfaction, and torment, those are what nourish science." The last remark is often heard in conjunction with the fine arts, i.e., that a little suffering helps to foster creativity.

Biologists must not remain complacent, and, judging from the vast amount of current literature, there *is* considerable anxiety and dissatisfaction with the way biological sciences are now being taught.

In his presidential address to the Linnean Society of London, C. F. A. Pantin (1962) discussed various approaches to the teaching of biology. He understood the necessity of maintaining close communication between the various subdivisions of biology and related fields of science in saying, "It is from these regions of overlap that the unforeseen scientific developments occur." More recently, Nils Sjoberg, Chairman of the National Committee for the Revision of Biological Instruction in the Gymnasia of Norway, was on our campus at Oregon State for several months as part of his tour in this country to study the changes occurring in high school biology teaching. I have encountered similar concern about biological instruction in The Netherlands and, at a recent United States–Japan symposium in Washington. Because science is universal, the problems associated with teaching science also seem to be universal.

Not many years ago, you would have encountered widespread uniformity in course content and course offering in biology. Today, you would find a bewildering variety of courses and curricula. Biological instruction today is undergoing considerable change.

The present state of biology is due in large part to its rapid postwar development. Jacob Bronowski, in a talk delivered in 1967, held that the impetus for this accelerated growth resulted, at least in part, from the fact that physical scientists and mathematicians turned their attention to biology after their intensive war-time efforts in their own fields. The expertise they brought with them rapidly opened new avenues of investigation. The result of this interdisciplinary research is, of course, well known.

Many equate modern biology with molecular biology, where fantastic strides have indeed been made, but other areas of biology have also advanced. Three examples will suffice to illustrate.

● Systematics has long been associated with the dusty confines of
museums, yet as I participated in the Systematics Institute, sponsored
by the Smithsonian Institution, the presentations of Robert R. Sokal
on "Numerical Taxonomy"; of Richard D. Alexander on "Animal
Behavior and Systematics"; of Charles G. Sibley on "Molecular Systematics" have long outpaced their traditional roles.

● Ecology has progressed far beyond the level of merely tabu-
lating floral and faunal inhabitants of field or forest. Examine if you
will the recent book edited by Kenneth E. F. Watt (1966) on systems
analysis and you will find one modern approach to ecology.

● Zoogeography achieved respectability in 1876 with the publica-
tion of *Geographical Distribution of Animals* by Alfred Russell
Wallace. Compare Wallace's book with the contribution by Robert B.
MacArthur and Edward O. Wilson entitled *The Theory of Island
Biogeography* (1967).

Biology courses have changed and so have the books used. As one
measure I compared the book I used at Iowa State in 1947 with *The
Science of Biology* (Weisz, 1967). As you might expect, the difference
was startling, not only in emphasis, but more than half of the information in the new book was simply not present in the older volume.
Clearly, we have come a long way in twenty years.

Ordinarily, scientists rarely complain about having too much information, but in this instance, the reception of modern biology has not
been uniformly enthusiastic. The major criticism seems to be that
"the whole organism is being ignored"—this at a time when we know
more about the whole organism than ever before. We now have a
much better idea of *how* it functions, when formerly we often knew
only that it *did* function. If we fail to utilize available knowledge, we
can only retard the growth of science. If we fail to familiarize our students with the latest advances in all aspects of biology, how can we
expect them to be prepared for the present, much less the future?
The most serious mistake we could make would be to educate our
students as we ourselves were educated.

The core curriculum has received considerable attention as one
means of updating biological education. It differs from past programs chiefly in emphasizing the unity of biology rather than the
differences. The Commission on Undergraduate Education in the
Biological Sciences has published an extensive analysis of four curricula considered representative of the major types of institutions
found in the United States, which should be studied by anyone in-

terested in this question. Cell biology, genetics, physiology and development receive much greater coverage than do morphology, growth and taxonomy. This is only to be expected, since most of the information contained in the first group has a direct bearing on the concepts presented in the second. I believe that a concerted effort is being made to relate the details to the whole. Though we have not achieved complete integration, an effort is being made in course and curricular design and in textbooks.

Many years ago we taught biology. Later, it was subdivided into botany and zoology and still later microbiology. Now we are back to biology again—not as a complete circle but more like a portion of a helix, since our level of knowledge is higher. The comparison with the DNA molecule is not mere accident, for surely the elucidation of the structure of this molecule ranks as one of the great discoveries in biology. Hegel's theory of historical development—thesis, antithesis and synthesis—is applicable, for modern biology is, in fact, a synthesis. As Whitehead (1929) applied Hegel's idea to general education, so well might we apply it to biological education.

One feature common to many core curricula is an integrated biology course rather than introductory courses in botany, microbiology and zoology. I believe this to be a distinct improvement for, as I have said earlier, it stresses unity. But it is not easy to design such a course. Right now the team of faculty members who will teach our course at Oregon State is deeply involved in deciding just how much time will be spent on what subjects. We have elected to offer our course during the sophomore year, preceded by a year of mathematics (normally calculus) and chemistry and with organic chemistry taken concurrently. Physics will come in the junior year. We are developing a second series of courses to follow the biology course during the junior year—genetics, cell physiology and ecology. Thus, our core will extend through most of the first three years for our biology students. Specialization becomes possible during part of the junior and all of the senior year.

There are other approaches, of course, but the goal remains the same—to provide the biology student with a sound education in the biological and physical sciences and in mathematics. All are necessary for fundamental work in biology. Biology has become one of the more demanding disciplines today because of this very need for expertise in related sciences.

How does this approach to biology affect students in agriculture or forestry? Are they any different from any other student of biology?

Some think so, but I do not. It has been said that to be an applied biologist, one must first be a biologist. I agree. To draw a line between applied and nonapplied biology is difficult, if not impossible. I would therefore urge serious consideration be given the biology core program. Some adjustments might be necessary, but I believe the effort would be worthwhile. The most obvious objection has to do with time—just how can a student meet the core requirements and still learn something about forestry or agriculture? One suggestion well worth consideration is a conjunctive tutorial section (or recitation), for credit, under the direction of a professor who is capable of relating the principles discussed in the basic science classes to problems of agriculture. If such a course were offered in parallel with biology, I believe the relevance of modern biology to agriculture and natural resources would become more apparent to students. It would be unreasonable to expect this to be accomplished within the confines of the biology course alone.

I oppose highly specialized programs at the undergraduate level. We seem to be enamored with formal course work. If a matter is worthy of consideration at all, we seem to think it merits at least one course and preferably a three-quarter sequence. This is quite the opposite of what I found to be true in The Netherlands, where I was much impressed by the ability of the laboratory staff. The educational backgrounds were amazingly uniform—a sound education in the basic sciences and training in how to apply this knowledge to the solution of agricultural problems. My colleague there had had but one formal course in entomology, and that of only four weeks' duration. Still, he had acquired a knowledge of the field equal to many entomologists I know who were educated under our system. He could readily enter adjacent fields in order to strengthen his research capacity. This ability to identify problems and bring diverse points of view to bear upon them must become the hallmark of our own students.

We are in a dilemma. How can we prepare students for the future when we do not know what the future will be? Their accomplishments will be determined by how well we do our job.

REFERENCES

MacArthur, R. H., and E. O. Wilson. 1967. The theory of island biogeography. Monographs in Population Biology, No. 1. Princeton University Press, Princeton, New Jersey.

McElheny, V. K. 1965. France considers significance of Nobel awards. Science 150:1013–1015.

Pantin, C. F. A. 1962. On teaching biology. Proc. Linnean Soc., London 173(1):1–8.

Wallace, A. R. 1876. The geographical distribution of animals. MacMillan and Co., London. 2 vol.

Walsh, J. 1968. Molecular biology research, Geneva. Science 195:718–721.

Watt, K. E. F. (ed.) 1966. Systems analysis in ecology. Academic Press, New York.

Whitehead, A. N. 1929. The aims of education. MacMillan Company, New York.

DAVID G. BARRY

Biology is one of the more rapidly changing fields in undergraduate education, in part because basic research has generated great volumes of information and in part from changes in the fundamental orientation of basic research. Similar issues confront engineering and agriculture.

The impact of the campus on society in the 19th century is not wholly clear. The scientist of that period viewed the natural world as static and unchanging. Our situation is now very different—views have changed and so have the responsibilities of the campus. Man's expanding knowledge of geology in the first half of the 19th century firmly established the concept of change, an exploration that culminated in the publication of the theory of organic evolution in 1859, events that modified our ideas about the origin of living things and of the earth itself. Then, in rapid succession, came the discovery of the atom, the electron, and x-radiation. Traditional concepts of matter were abandoned. The work of Hahn and Meitner finally brought us to the splitting of the atom and gave access to its enormous stores of energy.

The breakthroughs in molecular chemistry of genes and chromosomes and of enzyme energetics have changed the fundamental questions we ask of nature. A new relationship has developed between applied and basic biology. The applied scientist more than ever before faces the need to use new knowledge in directing the course of change in nature. It is our newly assumed responsibility to give direction to these changes.

The methods initially used in the teaching of biology reflected our concept of a changeless world. Students were presented with a mass of descriptive information called "the facts." In such an atmosphere, experimentation was the exception—all too frequently, the facts were taught as ends in themselves and drawing and observation were the order of the day. These methods were successful as far as they went and some very productive intellects were shaped, but new perspectives face our students.

Undergraduate biology today is under pressure from two forces. Graduate research continues to generate new information and concepts. The new high school programs continue to produce better trained students. We in the in-between areas are torn between our historic commitment to "cover the material" and our increasing realization that science must be taught as a dynamic intellectual process.

What is the contemporary state of biology? What are the great changes that have provoked us to talk about the "new biology," the new framework? We can describe biology as having passed through several phases:

- An early, natural history phase that emphasized the analysis of the distributions of species in space and time.
- A second phase emphasized analytic techniques. Out of this work came great advances in endocrinology and the understanding of hormonal controls, which defined the organism as a system with a multiplicity of feedback mechanisms and led us to awareness of the complex interrelationships between the individual and environment. It could be argued that at this time biology became interdisciplinary.
- A third phase brought a better understanding of metabolic processes. We began to appreciate the role of enzymes as specific catalysts, formulated symbolic models describing interrelationships between molecules and began to visualize form and function at the macromolecular level. These models represent abstractions that place new demands upon the student, requiring sophistication in mathematics and chemistry. New information at the molecular level provides answers to problems at other levels. There now exists a mass of information that far exceeds the limitations of our traditional approaches to undergraduate education.

In an educational pattern that depends more on analysis than on description, it is neither possible nor desirable to provide all the information that students must have in a single course. This situation leads directly to consideration of what should constitute a core of information for the undergraduate major.

Early in its existence, the Commission on Undergraduate Education in Biology decided to establish a Panel on Undergraduate Major Curricula that would make an "in depth" analysis of what was happening on several campuses that had undertaken extensive reviews of their curricula. It was decided that:

• Contact would be made with several top level institutions including those that made use of biology for professional school goals as well as basic science.

• That the details of the curricula would be reported in a form that allowed for analysis by computers.

• That the study should attempt to identify materials common to the different campuses, in the hope that the analysis would generate a body of information representing an agreed upon core of information to be presented to the undergraduate major.

In selecting the representative institutions, it was recognized that interest and willingness to cooperate were essential. The final list included: Purdue University (a large public university); Stanford University (a large private university); North Carolina State University (a large land-grant university), and Dartmouth College (a moderate-sized private college).

CUEBS staff representatives visited each of these campuses and sought to identify every item, concept or fact to which a professor allotted at least five minutes of time in the core program, on the assumption that a five-minute minimum would preclude those "merely passing references" that an individual instructor might make. Taking a 50-minute classroom period as a unit, five minutes represented one tenth of such a unit. The visitors analyzed the instructors' syllabi, lecture notes, student lecture notes, laboratory exercises and examinations. They reviewed lists of textbooks and assigned reading materials, but did not include them in vocabulary analysis. Through personal interviews, time allotments and value judgments were established, based on the information transmitted directly in lecture and laboratory. Thus, the analysis was designed to reflect direct teacher-student interchange, rather than information from supplementary sources.

The information bits (five-minute units) were recorded on I.B.M. punch cards that identified the item, the institution, the year, and the semester. The sequence of the information bits in a particular course, and the level of organization (molecular, cellular, zoological, botanical, etc.) was also established. Thirty-two hundred vocabulary items were generated in the four-year institutions, which was drawn into a common vocabulary for purposes of analysis.

Review of the assembled data suggests that, today, more than one year of general biology is required to prepare the student for advanced work. It shows further that the diverse specialties of contemporary biology are best served by a common introduction. Laboratory work that includes experience in experimental design is essential. While we have always assumed that the laboratories served at least four purposes—the transmission of information, the development of skills and techniques, the development of experience with experimental design, and the ability to carry out independent experimental design— we have yet to prove it. I submit that we have accomplished goals one and two rather well. We must now realize that we have never fully committed ourselves to items three and four.

Finally, it was concluded that analysis of the basic information bits should be made by vocabular categories, with topic and subtopic analysis. The results are of interest—any three out of the four institutions are in close agreement on each major vocabulary area. Agreement in the data and topics decreases as concern for detail increases, probably as a reflection of the tastes of the individual lecturer. As for the Dartmouth program, it also stems from the fact that their organismic biology is not a part of a required core sequence.

By definition, a core program is required. The greatest divergence in emphasis occurred in cell biology and genetics—relatively new areas of information—suggesting that teachers approach it from many facets and use a diversity of examples in representing basic principles.

Of course, a given information item could be classified under different headings, depending upon the proffesional tastes of the person performing the analysis. Thus a large fraction of the basic information can be considered as bearing on the molecular level of analysis, but also as related to genetics, to protein structure, to replication, to energy exchange, to basic physiology. The development of the vocabulary list was difficult, though performed by professional biologists in consultation with a number of their peers. In reviewing it one recognizes that the change in biology is primarily a change in how the information is organized and how the questions are phrased—it has been a quiet revolution at the fundamental level. There appears to be close agreement in program planning among the four institutions, although the programs were developed independently.

There is good agreement among the core programs as to molecular and cellular conceptualizations. These, when added to the concepts

of ecology and population biology, establish what can be called an
emergence discipline of general biology, dervied from general zoology
and general botany – biochemistry and biophysics. How much of the
detailed information is identical among the four institutions? Seven
percent (140) of the nearly 2,000 vocabular items are shared by the
four schools. This 7 percent, however, occupies about 16 percent of
the total time allotted to the core. If you consider any three of the
four institutions, then 25 percent of the 2,000 items account for
about half of the total instructional time (250 hours). This shows
good agreement among the different programs—but remember, it is
minimal—the similarity is undoubtedly greater than the vocabulary
list could demonstrate. The greatest commonality is in the field of
genetics, where one half of the items appear in all four institutions
at the molecular level; more than half are common to three out of
any four of the institutions.

What does this mean? One thing it surely means is that any curriculum that has not had careful faculty review and analysis within
the last five years is likely to be obsolete. This study should convince
faculty and administrative leaders of the importance of giving support
to continuing curricular analysis and study.

Just how close the curriculum should approach the research frontier is yours to decide. The new should not be adopted merely because it is new, but because it has significance. Each of the institutions
studied in 1965 has continued to modify its program; the survey is
already out of date as to details. There is need for a common core
approach that extends what used to be called introductory biology
into the second or even to the third year as is appropriate to schedules.

In my opinion morphology and systematics now become upper
level and graduate study areas that can employ sophisticated research
techniques. In no way has their importance been diminished, yet the
phylogenetic approach no longer appears to be a useful vehicle for
the dissemination of biological information, thus breaking a tradition
of some 2,000 years' standing. The contemporary biologist is concerned with operational concepts of the cell, development, and the
mechanisms of integration and evolutionary dynamics that lead to the
continuity in the levels of living organizations: molecular, cellular,
organismic, and population levels.

It also is clear that each institution must develop its own curriculum. There is no one "ideal" curriculum. A workable curriculum must
reflect the interests and competences of the faculty, as must the artic-

ulation patterns between two-year and four-year institutions. I recommend that you undertake a similar in-depth analysis of your own
curricula, to discover what you are actually teaching and to give you
evidence on where you should go. One prepares a student for an uncertain future by giving him a basic curriculum early—one that is
flexible and one that does not demand specilization too soon. Keep
in mind that 50 percent of all college graduates end up doing something that they were not formally prepared to do.

The core program—as distinct from the "core course"—should be
developed gradually, with participation by all who are teaching the
programs. There is no other way to plan an integrated presentation.
Efforts are being made to solve the problem of faculty obsolescence.

Finally, it is in this context that I recommend the following areas
be considered for inclusion in any basic biological program.

- Molecular basis of energetic synthesis and of metabolic controls.
- The nature of hereditary transmission of the basic properties of
cells.
- The function and development of major types of organisms at
different levels of complexity.
- The relationships of organisms, one to another, and to their en-
vironment, internal and external.
- The behavior of populations of organisms in relation to evolu-
tionary patterns.
- The bridge between the observable and the abstract.

Emphasis in different institutions will vary with the interests and
with the competences of individual faculties. But if we are aware of
how our teaching relates to the central body of contemporary bi-
logical thought and organize our programs accordingly, we will be
preparing our students as best we can. As professionals, this is our
fundamental responsibility to the teaching process.

G. FRED SOMERS

Traditional approaches to biology are being eroded away by fresh
steams of thought and experimentation. I propose to comment upon

the origin and impact of this erosion and to do some guessing about the future.

From whence comes the current ferment? Why has re-examination of our curricula in biology become desirable? What have efforts at reassessment produced? What does the future hold in store?

For decades the philosophic focus of biology was upon the theory of evolution. New findings were, for the most part, tested against this working hypothesis. Genetics made possible some understanding of evolution as a vital process, but even here progress was slow.

Then following the announcement in the early 1950's of the Watson-Crick hypothesis of the structure and role of DNA, research in biology underwent a sudden shift in focus. A whole new framework of theory was now available, which researchers attacked with enthusiasm and vigor. In the process, the boundaries between biology, chemistry, physics and mathematics became even less distinct than they had been in the past. But the teaching of biology remained essentially unchanged.

In the late 1950's a number of biologists became concerned over the growing hiatus between teaching and research in biology and organized several conferences, the summaries of which were published in 1958 (Committee on Educational Policies, 1958). There the matter rested for the most part. Little happened until the Commission on Undergraduate Education in Biological Sciences was organized in 1963. Out of meetings sponsored by this latter group there grew the concept that there is a central body of knowledge with which all majors in biology should become familiar. To this was applied the terms core program or core curriculum.

The validity of this notion has been examined in some depth by conferences at state, regional and national levels. The Commission has since established study panels to examine various more restricted aspects of the core program concept and the unavoidable problems that attend its introduction and a consultant bureau to provide expert assistance in planning revised curricula or new facilities.

You should know something of the contents of Commission publication No. 18 entitled *Content of Core Curricula in Biology. Report of the Panel on Undergraduate Major Curricula* (Grobstein *et al.*, 1967).

In the first place the term "core curriculum" does not mean the same thing to everyone. To some it means a closely knit, well-integrated, coherent sequence of courses, often necessitating the abandonment of essentially all current courses—a very expensive and time-consuming venture. Others, recognizing that sweeping innovations are very difficult to accomplish, took a different route. They assembled a melange of traditional courses into a package that they termed a core curriculum, although those who have given most thought to the philosophical bases for a core program deem this maneuver less than satisfactory. They consider it only an intermediate step toward what must inevitably be done eventually. Still others, seemingly satisfied with their present situation, largely ignore all these developments.

While the Commission recognized that there are obstacles associated with the development of a core program, they rejected the notion that the Commission should itself develop a "model program" that might, in a sense, be sold to biology departments across the country. They felt that this was *not* their role, but rather that they should serve to stimulate groups at various universities to examine their objectives, to deal with the problems that were indigenous to their circumstances and to arrive at a solution that was viable for them.

No two situations are the same. In some cases there are already in existence closely-knit, well-integrated departments of biology. In other cases biology is represented by a multitude of departments, sometimes representing two or more colleges. To integrate the latter group represents a herculean task for which leadership is difficult to find and involves mammoth problems of compromise and salesmanship. Indeed, where this situation exists, the attainment of a close-knit, well-integrated core program in biology has been difficult to attain and it may take decades before it can be accomplished. In some cases integration can be achieved by forming colleges or divisions of biological sciences—in very large institutions this may be the most practical solution.

The Commission assigned to its Panel on Undergraduate Major Curricula the task of defining the specific content of existing core curricula. The Panel's strategy was as follows:

● Select four high-quality but rather differing institutions that had recently given serious attention to the content of their biology curricula. Those selected were Purdue, Stanford, North Carolina and Dartmouth.

● Record the curricula in sufficient depth and detail so as to enable them to be analyzed and compared.

● Identify the common materials and organize them in a form permitting effective communication with other interested institutions.

This study is reported in CUEBS publication No. 18 (1967). Examine carefully the basic assumptions and the methods used by the Panel, for while the report has limitations: It is the most comprehensive study available to date. It shows the measure of agreement amongst four institutions, whose programs have served as models for others—one of which, at Delaware, is summarized in the accompanying table.

When the content of the curricula at Purdue, Stanford, North Carolina State and Dartmouth are compared, there is a surprising degree of unanimity—relatively great emphasis on cell biology, followed by genetics, physiology and development. Rather trivial emphasis is accorded such items as evolution, ecology, growth, and tax-

Basic Requirements for Biology Majors at the University of Delaware

Year	Biology	Chemistry	Mathematics	Physics
Freshman				
1 Sem		General	Introduction to calculus	
2 Sem		General	Introduction to calculus	
		Qualitative analysis		
Sophomore				
1 Sem	Concepts	Organic / Organic Prep.		
2 Sem	Developmental	Organic / Organic Prep.		
Junior				
1 Sem	Cellular and molecular / Genetic and evolutionary			General
2 Sem	Environmental			General
Senior				
1 Sem		Physical		
2 Sem	Senior Seminar	Physical		

onomy. In the past, evolution and morphology would surely have played a much more dramatic role in a biology program, an aspect that reflects a rather profound change in what is considered desirable for undergraduate biology.

A similar, though less pervasive unanimity appears in the "topics in cell biology," but with wide divergence in such topics as enzymes. "Topics in evolution" are allotted only 0–4 percent of the curriculum content—a very small amount of time, relatively speaking, to topics that would have been major items in the past.

When one examines these core curricula in terms of "levels of biological organization," one once again finds a considerable amount of agreement. Cellular and molecular levels of organization play a much more prominent role than they have in the past, but consideration at the organismal level predominates. Topics at the population level are apparently left to more advanced courses.

Finally, if one looks at "major biological disciplines," one finds that about one third of the time is devoted to "general biology." By this the Panel obviously has reference to topics that are not clearly related to a single phylum—there is, however, a rather good balance amongst the phyla. To some extent it is a bit surprising that microbiology does not show up more strongly, in view of the fact that many of the people engaged in development of core curricula are doing research in this area. Perhaps it is included under the term "general biology."

Clearly, we are not viewing an accomplishment, but a process. Biologists are finding the necessity and opportunity to examine their teaching. There are a number of compelling reasons why the process should continue:

- one faces an ever-increasing number of students, staffing is becoming more difficult, costs are soaring. At the same time new technologies and a wide range of sophisticated new gadgets are available.
- There are many who feel that biology needs to be treated more as an entity, that we should emphasize *biology* and not botany, bacteriology, and zoology. They suggest that the content of biological sciences might better be organized according to functional levels— molecular biology, cellular biology, developmental biology, organismic biology, genetic biology, population biology. There are various modifications of this theme, but all of them differ markedly from the traditional pattern.
- Mathematics, chemistry and physics have increasing impact

upon the research approaches being used. The tools of the physical sciences and mathematics are largely insensitive to taxonomic hierarchies; it is more efficient to use them without regard to such boundaries. If students are to include such approaches in their schooling, something has to give. There are those who feel that the taxonomic-phylogenetic categorization is in a measure an anachronism, no longer meriting the attention it once enjoyed.

● Possibly, out of a new approach to the subject will grow new generalizations. This *could* be the most important reason of all for redirecting our attention to the organization of knowledge in the biological sciences.

REFERENCES

Commission on Undergraduate Education in the Biological Sciences. 1967. Content of core curricula in biology. Report of the panel on undergraduate major curricula. Commission on Undergraduate Education in the Biological Sciences, Washington, D.C., Publ. 18. 176 p.

Committee on Educational Policies, National Research Council. 1958. Recommendations on undergraduate curricula in the biological sciences NAS–NRC Publ. 578. National Academy of Sciences, Washington, D.C. 86 p.

ROBERT H. BURRIS

It is always hazardous to predict the future in science; we usually prove much too conservative in our estimates. From a careful examination of the past, and an inspection of recent trends, we try to extrapolate to the years ahead.

One obvious trend is that biology has become more quantitative in recent years. Gross description has made its contributions and needs to offer no apologies, but it is a great mistake to insist on continuing emphasis on a field that is clearly declining. We must move ahead, and as a result certain biological disciplines may suffer. Of course, there is still work to be done in descriptive biology, but it would be dishonest to persuade college students that this is an area of major challenge for the future. Young people are perceptive, and they will not move into areas that they recognize as unlikely to contribute substantially in the future.

Many descriptive biologists have switched their emphasis to ultrastructure. Commercially available, high-quality electron microscopes have opened up new dimensions and have revealed many fascinating details about the substructures of the cell. Numerous laboratories are exploiting electron microscopy to define structure that can be correlated with cell function. It now is apparent that many of the subcellular bodies serve as compartments for specialized activities in cellular metabolism.

Systematics has also been de-emphasized in the general movement toward quantitative biology. Investigative approaches are being modified, and many systematists are adopting powerful chemical and biochemical tools, such as thin-layer chromatography and gas chromatography, as aids in classification. These methods must be used with caution, but they are tools of great promise in distinguishing between closely allied species.

As a corollary to the observation that biology has become more quantitative, it is apparent that the training of biologists increasingly must emphasize mathematics, physics, chemistry, and biochemistry. These are the disciplines that are of greatest help in studies of detailed biological systems.

A second trend-maker that emerges is a general recognition that there is unity among biological systems. The past witnessed a movement that divided biology into many subdisciplines. Now the movement is in the other direction, and virtually all units set up in new universities are pooled under the title of biology to embrace botany and zoology and often microbiology, genetics and biochemistry as well. Fusion of established departments may be a traumatic experience, but, in general, it is desirable to join forces and to attack biological problems on as broad a front as possible.

There is an inherent unity among the biological reactions of plants, animals and microorganisms. To be sure, there are certain specialized reactions in each of these groups, but basically the biochemical reactions and the cellular functions are remarkably similar. Comparative biology has been taught for generations, but there clearly is a greater appreciation now of what comparative biology can teach us than there was formerly. Investigations of basic reactions have shown that the mode of action of enzymes differs only in minor details from one organism to the next. We are just beginning an era in which establishment of the tertiary structures of proteins will form a rational basis for explaining the enzymatic and immunological activity of proteins.

Spectacular advances in genetics have made it one of the most productive and promising areas in biology. It is a relatively new discipline, but attacks on problems at the molecular level have provided insight into some of the most fundamental of life processes. Molecular biologists to date have largely emphasized problems in genetics. The discovery that DNA is the material that transfers genetic information, the establishment of the DNA structure, and the demonstration of the transfer of information from DNA to RNA and thence to protein, have provided insight into information preservation, duplication and transfer. We quickly reached the point where it was possible to achieve the chemical or enzymatic synthesis of polynucleotides in a defined pattern. Transfer RNA has been crystallized in several laboratories and now is amenable to x-ray analysis; knowledge of its tertiary structure will follow and should be crucial in establishing its detailed mode of action. The genetic code has been unraveled, and the function of each of the 64 triplets in the code has been assigned; there is much information on how the codons define protein synthesis and how synthesis is initiated, terminated and otherwise controlled. The possibility of controlled modification of the information on the gene is close to being realized, and the completely defined synthesis of a substantial gene unit has been achieved. In short, our knowledge of genetics has exploded during the past two decades, and the promises of its applications in the future are overwhelming.

It is apparent that much of this new information will be utilized in a practical way for modifying plants and microorganisms. Manipulation of genetic material in the past has tremendously increased the productivity of plants and animals. Now that genetics is better understood, its implications for agricultural productivity promise more spectaculars comparable to hybrid corn and high-lysine corn.

New experimental tools have greatly expanded potential biological research, although some biologists have a tendency to avoid sophisticated instrumentation and to take refuge in traditional descriptive approaches. This escape mechanism is hardly justifiable; a biologist should utilize every tool that can help him achieve better insight into the problems he is studying. He may have to go out of his way to borrow the tools or to seek collaboration in their use, but he should do this whenever necessary.

For example, chromatographic methods are a powerful means for the separation and identification of compounds in low concentrations. Even with simple equipment, chromatography allows the investigator to make separations that would have been virtually impossible or prohibitively time-consuming 25 years ago.

The mass spectrometer and nuclear magnetic resonance provide means for establishing organic structures in a fraction of the time formerly required. Not only is the identification speeded tremendously, but characterization can be done on much less material than is needed for conventional organic analysis.

Visible, ultraviolet and infrared spectrometry give a great deal of information about complex compounds. Free radicals can be detected by electron paramagnetic resonance. Circular dichroism and optical rotatory dispersion give information on the configuration of proteins. Molecular weights of macromolecules can be established with facility by ultracentrifugation or by separation, together with standard reference compounds, on gel columns. Isoelectric focusing furnishes much more accurate information on the isoelectric points of proteins than has been available previously.

Instruments are being built so that the data they gather can be converted directly from analog to digital form for computer analysis. Furthermore the computers can be connected to animals or to isolated organs so that a change in the subject will elicit a specific computer signal that is fed back immediately as a stimulus. Biologists have only begun to utilize research tools effectively, and they can expect to obtain information more rapidly and accurately by applying the increasingly sophisticated instruments that are becoming generally available.

Lest the implication remain that all is now being run by computers and by automated instruments, note that there is increasing appreciation of ecology and its potential contributions to biological sciences. The ecologist has long been a voice crying in the wilderness, a voice coming through with increasing clarity as man continues to degrade his environment by pollution of the air, the water and the land surrounding him. Desecration of our environment already has reached alarming proportions, and the biological balance that we have known is being rapidly destroyed. Somehow we must learn to respect the environment and to live in harmony with it, and the ecologist has much to tell us about developing a rational program.

The interacting factors in ecological systems are extremely complex. It is no longer adequate to be merely descriptive; again the information must be quantified. This will require extensive analyses of pollutants, development of methods to minimize pollution and detailed studies of the interaction of organisms among themselves and with their total environment. The words of the ecologist are being heeded, but he must be given support by those who command diverse skills to aid him in a rational study of environmental problems.

Knowledge now has advanced to a point where new emphasis and more meaningful research can be applied to multitudes of old problems. For example, the biologist has long been concerned with developmental biology and the control of biological reactions, but at the descriptive level. Now methods are available that may explain in detail how development occurs in the young plant or animal. An obvious correlative to studies in development are those of control mechanisms, for development is in essence a series of controls and relaxations of controls at suitable times in the developmental process. Product control, feedback control, genetic control and hormonal control are all targets for detailed study.

Neurobiology in the past has told us much concerning the chemical reactions that occur at the synapse and has given much information about the electrical nature of nerve processes. Now the methodology and background available to the investigator have improved to a point where we can expect rapid advances in some of the more challenging problems in neurobiology, not only in signal initiation and sensory transfer of information but also in the processes that occur in the brain itself. Imaginative scientists are attacking problems of neurobiology from many angles, and there is promise that major advances will be made in a relatively short time.

The importance of membranes in separating the contents of the cell from an unfavorable external environment has long been appreciated, but there has been less awareness of the role of the intracellular membranes and their function in compartmentalizing various activities within the cell. Biological membranes are marvelous structures that exclude certain materials and allow others to pass selectively, a selectivity that is under metabolic control. The electron microscope has uncovered much of the ultrastructure of these membranes, and many aspects of their chemistry have been revealed. There are still many questions concerning how membranes select one material and exclude another—far from being passive barriers, membranes are complex and subtle.

Immunology is still another area that has advanced from an empirical approach to an approach that has a rational basis. Investigation of the detailed structure of proteins will tell us more and more about how antigens and antibodies interact and the nature of the specificity of these reactions. Accurate knowledge of immunology becomes increasingly important as it relates to the transplantation of complex organs. Rejection of the organs can be blocked effectively only on the basis of a thorough knowledge of immunological reactions and with due regard for the reduction in host resistance to invading organisms.

Biophysicists are active in applying the physicist's approach to complex biological systems. They have introduced sophisticated electrical and optical instrumentation and have helped define the tertiary structures of several proteins by x-ray analysis, studies that are fundamental to our understanding of protein and enzymatic reactions.

If any one thing comes through crystal clear from an examination of the advances of biology in the past two decades, it is that the rapid developments in biology were largely unpredicted by those who taught in an earlier age. Because the advances were unpredicted, it follows that the most pertinent part of our biological training was the block of basic principles that we learned. It is manifestly impossible to cover all aspects of biology in any training program, and over-concentration on biological details of current interest will be relatively meaningless in the future. It is incumbent on teachers of biology to give each student a solid background of biological fundamentals that he can later apply to a variety of problems. Only by emphasizing fundamentals can we give a student the flexibility to adjust to changing times and new information.

The realization that we must emphasize basic information in biology led logically to the concept of a core curriculum. I view this not as a currently popular gimmick but as a fundamentally correct approach to the teaching of biology. Specialized courses must be de-emphasized and the basic core courses stressed. It is encouraging to find that the response of students to core curricula is generally favorable, probably due in part to a fundamental validity and in part to the fact that they are new and hence inherently challenging.

The instructor is obliged to add a touch of spice to the fundamentals, perhaps by illustrating the fundamentals with exciting contemporary examples of specialized biology. Interest can also be created through special lectures that explore biological investigations into diverse fields. Small seminar groups also can probe biology in a fashion that will maintain a high pitch of student interest.

There are certain pitfalls in instituting core curricula. One is the obvious temptation merely to reshuffle existing courses into a new format and then to declare that a new core program has been created. Quite the contrary, a core program should develop from a detailed discussion of students' needs, a discussion including representatives not only from biology, but also from chemistry, physics, mathematics, biochemistry, genetics and from other departments that can contribute. Transition to a core curriculum provides an excellent opportunity to re-examine critically the traditional approaches to biology teaching. Above all, it offers chance to reorganize material in a fashion that will permit better integration of concepts.

A second hazard inherent in the core curriculum arises from its supposedly new and modern image. Awed by this image, the instructor may feel obliged to present only material that appeared in journals published during the preceding month. True, some of this material may furnish pertinent illustrations, but the core curriculum should emphasize fundamentals, should indicate the unity of biological systems, and should not be embarrassed to draw on the past for valid information.

The biology curriculum should not overemphasize biological subjects at the expense of other topics. It becomes increasingly important that every trained biologist have a good background in chemistry, including organic and physical chemistry. He should pursue mathematics at least to the level of calculus, and he should be given an opportunity to study computer science and statistics. A year of physics should be obligatory. A course in biochemistry is helpful to bridge the gap between chemical and biological principles.

Attention has been directed here primarily to the student who wishes to emphasize biological science in contrast to agricultural production. Many students who specialize in agricultural science go on to graduate school; emphasis on a core biological curriculum and on other basic sciences will fit an agricultural student admirably for advanced graduate studies. Indeed, what is good for the biology major is good for the agricultural scientist who wishes to specialize in microbiology, biochemistry, genetics, plant pathology or many of the other agricultural sciences. As in engineering, there has been a swing in agricultural schools to an emphasis on principles. The number of people directly employed in agricultural production has been decreasing, but at the same time there has been an expansion in the agricultural sciences that make the high productivity of our farms possible.

The student of agricultural science does not need a core curriculum designed specifically for agriculture. He needs the same basic information and basic research techniques as do other students in biology—the applications to agricultural problems will appear as he engages in research. Basic problems doubtless will be suggested by practical observations. The differing goals of agricultural production and agricultural science should be kept clear—they require a different type of training. The core curriculum in biology is entirely appropriate for the agricultural scientist; it is acceptable, but probably not ideal, for the student in agricultural production.

Cellular and molecular biology have made great strides and will remain areas of heavy emphasis. Investigation will be concerned with the nature of genetic processes at the molecular level. A new dimension has been introduced with the crystallization of transfer RNA, for this opens the possibility of detailed x-ray analysis of the tertiary structure of nucleic acids. The tertiary structure in turn should give further insight into the functioning of genetic materials that transfer information within the cell. Genetic information is basic to the formation of functioning proteins, and an understanding of the details of protein operation will be possible only when we know the tertiary structure of these proteins as established by x-ray crystallography. Enzymatic catalysis and its modification by changes in the chemical and physical environment will be appreciated better when the tertiary structure of the enzymes is understood. Immunological responses fall into the same category, because these responses depend upon the interactions of proteins; subtle immunological responses fall into the same category, because these responses depend upon the interactions of proteins; subtle immunological changes will be interpretable only when the exact structure of the antigen and antibody are established. x-Ray analysis of the noncatalytic structural proteins should reveal how they form the framework of the cell and its subcellular units.

In organismal biology, the biologist will continue to ask how cells function in concert, and how organs function together in the intact organism. Organismal biology stresses problems of organization and control. In the earliest stages one is concerned with genetic control of the fundamental properties of the organism and its parts. Then there must be concern with cellular interactions, because metabolic products of one cell may have a dramatic effect upon the functioning of other cells. By the elaboration of hormones one organ can control the function of another.

Much remains to be learned about control through the nervous system. One can seek data on initiation of signals and the transmission of these signals from point to point and on the nature of the receiving stimulus in the sensory process, both at primitive and at highly sophisticated levels. The biologist also will be greatly concerned about the still mysterious processes of memory storage and recall. He must explain how the stored information interacts to synthesize concepts from the various centers of memory.

Obviously, the ecological approach is highly complex and requires an improved definition of its problems and a better quantitation of its results. Much of the data collected will be of such complexity that it will require computer analysis to establish trends. From studies of ecology should emerge a better appreciation and respect for our environment. Since the agricultural scientist is basically attracted to the land and respects it, he is particularly concerned with ecological research. Colleges of agriculture present an especially logical setting for research in environmental sciences.

In general, specialized courses take care of themselves. They tend to proliferate, because they are pushed by people with a special interest and motivation, and should therefore be restricted in numbers. They must be founded on a solid base of chemistry, physics, mathematics and fundamental biology. If the student is to get an adequate background in science as an undergraduate, he has relatively little time for specialized courses; however, a student with proper basic training can manage the specialties.

No one would dispute the view that biology is basic to agricultural science. This is an era in which biology is in the ascendancy, for biology holds a multitude of exciting challenges that we dare not ignore. We should take a hard look at our agricultural curricula and replace that which is obsolete with biological training that is basic, valid and challenging.

N. N. WINSTEAD

Within a university there are many policies and interests that affect curricula, the individual teacher, and his course. Of these, two merit attention.

The first is concerned with our faculty reward system, where for a number of years we have witnessed a growing emphasis on research. Unfortunately, this emphasis has occurred at the expense of teaching —especially in the teaching of undergraduates. National prestige in a given discipline is associated almost entirely with research productivity, while undergraduate teaching has just not provided sufficient prestige and reward for the individual faculty member—this last is a concern on almost every university campus. Today, universities are attempting to re-emphasize the importance of teaching. One technique, imperfect though it be, is to supplement the usual university sources of information with data derived from student evaluations of faculty and from alumni surveys to help identify faculty who are considered good teachers— and then to make certain that the information is taken into account when salary increase and promotion time comes around.

The second item is concerned with interdisciplinary problem solving, whereby the development of interdisciplinary research programs has taught us that faculty from different departments, and even different schools, can work together—and that in working together, can solve the big problems. For example, ecologists are found in a tremendous number of departments—in biological and agricultural engineering, economics, and sociology, as well as in crop science, botany, plant pathology, entomology, microbiology, soils, horticulture, zoology, forestry, and wildlife. Ecology, in turn, is made up of a multiplicity of subecological areas, such as micro-, macro-, general, population, biomathematical, marine, classical, physiological, crop, plant, animal, and wildlife. In using ecology as an example, I do not suggest that we have too many ecologists.

We are beginning to see biologists from different departments—or from different schools—teaching and developing curricula jointly. At North Carolina State, ecologists are teaching courses together and even working toward a joint graduate degree program. Faculty from the departments of soil science and zoology in the School of Agriculture and Life Sciences and from the Department of Forestry in the School of Forest Resources developed a curriculum in conservation— which is, of course, a biology curriculum. While it may be inaccurate to say that a trend exists in the area of teaching, we are at least seeing many more interdisciplinary efforts.

University administrators have become much troubled by general proliferation of courses, and with the real question of effectiveness, as well as with the rising costs of courses and curricula. For example, North Carolina State offers 2,000 courses. At the recent conference

on biological undergraduate curricula needs for various agricultural
and natural resource areas, there was almost unanimous recognition
that today's curricula need to undergo a general overhauling and that
the three groups—biology, agriculture, and the natural resources—
must work together to meet their curricula needs. Many excellent
ideas have been generated, but one must recognize that the cooperative revamping of curricula will be far more difficult than it was to
bring together research components into desired groupings. This may
be due mostly to the limited availability of grant resources for instructional as distinct from research purposes.

One should hardly expect botany and zoology—the traditional
teaching departments—to do all the teaching in the biology core instructional program. Why should not the faculty in agriculture, in
natural resources, and in biology share the load? Certain areas in core
programs can use team-teaching—introductory biology, cell biology,
ecology, molecular biology, and developmental biology. While some
progress toward using interschool faculty has been made at many universities, much more improvement is needed.

If we are moving to more interdisciplinary courses, will that reduce
the total number of courses? All can agree that it should—but few
feel that it will. Apparently the present trend toward proliferation
of courses will continue.

There is a definite trend now to require fewer hours and courses
for graduation—coincident with this is a general loosening-up of curricula, so that students have more freedom to choose courses that
they wish to take. Freedom of choice occurs primarily among the
electives and general education requirements; majors continue to be
structured more rigidly. In truth, this "choice system" sounds better
than it works. To avoid forcing students into the same mold and to
provide them with a chance for individual development are goals that
we seek—yet we expect a graduate to have at least minimum competence in a given field. Reduction in total hours provides more time for
the "hidden curriculum" that some insist is the most important component of a student's education. There are arguments pro and con on
the details of balance and freedom, yet I suspect that most of us
would agree that the trend to more freedom is desirable. In any event,
curricula committees must take a hard look at the increasing number
of courses taught and start weeding out those least needed. Faced by
the trend toward requiring fewer hours for graduation, departments
sometimes react by dropping service courses. Can biology departments afford to do this?

Are curricula geared too much toward the training of students for graduate school? After all, biology curricula were in many instances intentionally developed as preparatory to graduate school. The idea was that terminal students would shift into curricula with vocational emphases. But in looking at the curricula and especially at the syllabi from many departments, one wonders what is happening to the curricula designed for the students who do not wish to obtain a Ph.D. Are too many undergraduate curricula designed primarily for the preparation of graduate students? This seems to be a trend, yet many students will not pursue and obtain a Ph.D.

With the trend toward heavy emphasis on molecular biology, introductory courses have been developed in many universities that have molecular biology as the central theme. At the same time there is a tendency to use a single introductory course both as a service course for the arts and sciences students and as a core course for the biology and agriculture students. I do not know which trend is better, but I suspect that molecular biology would be an almost total failure as the service course for nonscience majors, and that it may not be the best course for freshmen at a large number of our universities.

At Raleigh we have held tenaciously to a one-semester introductory course in the core curriculum that also serves as a general education course. We try to show in this course that diversity is but a "variation on a theme." Hopefully, this will encourage the student to see the unity of life rather than the multitude of confusing facts associated with an array of diversified forms. I prefer this approach but I must admit that we seem to be moving against the national trend by retaining a one-semester introductory course.

Entering students vary greatly in their competency and background in biology. Some have had excellent biology courses in high school—others still think that respiration means breathing. While advanced placement applies to a few students, one still wonders whether all 750 students in the "BS-100" really belong in the same course. Is there not some better education device than having one class of 400, one class of 350, and 25 lab sections of 30 each? I believe we must use advanced placement more frequently and provide for credit-by-examination to avoid having students repeat what they have already learned. Perhaps one technique to encourage more students to take advanced placement would be to give them an A or B, plus four hours of credit.

Another question continually arises, i.e., are laboratories necessary? With the increasing size of classes there comes increasing pressure to drop laboratories from science courses, for obvious reasons. Last year, several of our most distinguished professors at Raleigh led small group discussions for nonscience majors in lieu of lab in freshman biology on subjects such as pollution, population, environment, and food. The students, incidentally, considered this trial an immense success. Perhaps some of the labs now taught in agriculture, natural resources, and in biology should indeed be dropped, permitting us to devote our inadequate resources to develop first-rate laboratories in a smaller number of courses. At the same time, I do fear that we are about to drop the only remaining lab required of the humanities and social science majors.

In spite of all that we read, day in and day out, about automation, audiotutorial techniques, programmed texts, computer-assisted instruction and independent study, they have not in fact arrived on most campuses. The hardware is not the only holdup—although it serves as a good excuse and is very expensive. The software is sadly deficient! Much more time and effort is required of the faculty members under these conditions than when a course is taught in the regular way. Some day mechanical media may become established, but based on our present pace, most of us will apparently not live to see the day when they play a significant role in biology instruction except in isolated cases.

Graduate students continue as instructors of undergraduates. We do see some evidence of better coordination and training of these graduate student teachers to assist them in handling laboratory sections or, at times, lectures. Yet, there is a real need for more emphasis in training graduate students as teachers. Unfamiliarity with mundane things—peculiar or old-fashioned projectors—and even more importantly, lack of familiarity with new and complicated teaching media, mechanical devices and instruction techniques is an impediment to the new teacher's effectiveness. Just how does one use such gadgets and software effectively? We can help teaching assistants to become more effective and less frustrated by having a competent, experienced, and excellent teacher work with them. One can only hope that our senior and most distinguished faculty will continue to teach undergraduate courses. Should we not put our heads together and come up with a better understanding of the most effective roles for faculty and graduate students in biology instruction?

A related question has to do with need for coordination among courses and instructors—not only in the core biology curriculum, but also among the core courses and the courses that require them as prerequisites. To cite one example, a few years ago we had a delightful experience at Raleigh—many said it would be traumatic because it came during the Christmas holidays—at which each teacher in the core curriculum and a few related courses gave a syllabus of subjects covered in his course to others participating in the discussion. Then as we discussed the content of each course, we found, for example, Mendelian genetics and the Kreb cycle covered an embarrassingly large number of times.

Hopefully, a trend will emerge whereby teachers will know what is taught in the prerequisite courses, where only essential prerequisites will be required and where sequential courses will be built upon the prerequisite. But we ought to ask more questions: Does the undergraduate student really need a course in molecular biology before he can take ecology? Does Introduction to Horticultural Crops build on BS-100? How much more can be covered in ZO-300 if the course content really begins at the point to which the introductory course has taken the student?

It should be evident by now that I do not distinguish biology from agriculture and natural resources. I see them as components of the same program. I am encouraged as I look at the situation on the Raleigh campus. Best of all, our faculty is starting to ask the hard questions.

Although the decline of the classroom has been predicted by 1980, I think that prediction is wrong. I do think that libraries, specialized programs and devices, and independent study should and will play a more important role. Yet, I do not see these as threats to the classroom teacher. The teacher-student contact will continue to be the most important means of educating future students. Devices should be used to free the teacher and the student from busywork and from that portion of the learning experience best accomplished in other ways, thus enabling the teacher and student to enjoy a more effective and enjoyable interaction.

4

Physical Sciences and Mathematics

RICHARD M. SWENSON

Although the Commission's committees on chemistry, physics and mathematics operated independently, all three arrived at several similar conclusions and recommendations:

- A recognition of the rapid changes that have occurred and are occurring in agriculture and natural resources, and of the complex task of preparing graduates for these moving targets
- A recognition of the equally rapid changes that are occurring in the basic sciences, and the ever-increasing need for agriculture and natural resources students to have a sophisticated understanding of the physical and mathematical sciences.
- That specially-designed courses be neither watered-down versions of regular courses, nor survey courses, nor terminal in nature and that they encompass a clientele broader than just agriculture and natural resources—biology majors, premedical students, etc.
- That the background of the teacher and his ability and willingness to use appropriate examples oriented toward agriculture and the

71

biological sciences are of great importance. As the Committee on Mathematics put it: "Some teachers of mathematics must be encouraged to gain insight about agriculture and natural resources so that they can make mathematics a living and significant subject to students in the field."

• That there is critical need for a group to prepare relevant and appropriate source materials that professors may use in orienting their courses to agriculture and the biological sciences.

• That more time should be devoted to basic science and less to courses of the "how to do it" variety.

COMMITTEE ON CHEMISTRY

The Committee on Chemistry recommended a minimum of one full academic year (10 semester hours) for all students in agriculture and natural resources. This course should clearly show the impact of modern chemistry on society and should present an adequate overview of chemistry. It was suggested that it be comprised of 90–100 lectures, 30 discussion sessions, and 30 laboratory periods. It would involve integration of inorganic, physical, analytical, and organic chemistry. The exact order of topics may vary appreciably, but the organic and biochemical components should come early in the course so that examples involving organic molecules can be incorporated.

The Committee felt that this course represents the barest minimum and that only a very few agriculture and natural resource majors have this as their sole requirement. Serious consideration should be given to adding a second course as the eventual minimum for all students in agriculture and natural resources. Certain majors will require even more chemistry.

It was recommended that the second course be regarded as a continuation of the organic, biochemical, and analytical components of the first course and that it comprise about eight semester credits. Thus, students who took both courses would have completed the equivalent of:

• A standard one-semester course in general and inorganic chemistry.
• A one-semester course in qualitative and quantitative analysis.
• Sufficient physical chemistry to serve as a basis for the analytical and biochemical components.
• An introduction to biochemistry.

These courses should not be confused with survey courses but, as the Committee report stated, "provide appropriate depth for all those who need to make specific use of chemical concepts later in their careers. Indeed the Committee is of the opinion that they would not be inappropriate as the starting point in the professional training of chemists, which at the present is in danger of becoming too narrowly channelled."

From this point, the Committee felt, "the student can readily move into intermediate and advanced courses in chemistry as he needs them after taking the appropriate prerequisites in physics and mathematics."

COMMITTEE ON PHYSICS

The Committee on Physics surveyed 73 colleges of agriculture and found that some programs (in agriculture and natural resources) require no physics at all—most require from one quarter to one year, few require in excess of one year. The Committee concluded, further, there was but little enthusiasm for the courses now offered, a wish that they could be somehow different, more practical, more oriented toward the interest of the agricultural scientist, and more sympathetically presented.

It was the conviction of the physics committee "that almost every agriculture student should experience during his undergraduate training at least a one-year physics course that is more than a watered-down first course in physics for students planning to become physicists or engineers." They base their conclusion on the following reasons:

• More and more aspects of agriculture are emerging as quantitative.
• The widespread and rapidly-expanding accessibility of computers makes familiarity with them highly desirable.
• Research fields and graduate education in agriculture and natural resources demand a more sophisticated background in physics than it did formerly.

It was pointed out that the mathematics preparation now offered to precollege students makes feasible study of physics by more college students. Most applications of physics that are at the forefront of present agriculture science are accessible to any student with a working understanding of ordinary calculus and differential equations with perhaps a little introduction to modern algebra, probability theory, and rudimentary computer training.

The backbone of the recommended course should be fundamental physics, with a shift of emphasis as to illustrative material. Rigid body mechanics should be played down in favor of elementary fluid mechanics; the idea of thermodynamics should be emphasized more than statistical theory of heat; radiation laws should be treated empirically rather than derived theoretically, so they could be introduced at a more elementary level.

Laboratory work should include exercises with computers and perhaps some biophysical and meteorological measurements. Concepts of elementary calculus should be employed regularly.

COMMITTEE ON MATHEMATICS

The Committee on Mathematics placed special emphasis on the mathematical requirements of students in agriculture and natural resources ten or fifteen years from now. It pointed out that, even now, the application of mathematics here is highly sophisticated and for this reason they recommend requirements that are heavy compared to those now imposed. The Committee further emphasized that the spectrum of mathematical topics now beginning to be used in agriculture and natural resources is broader than in any other field except, possibly, engineering.

The following schedules of mathematical instruction were recommended as goals to be attained in the next ten to fifteen years:

Course Name	Recommended for Curricula in		
(Semester Hours)	Education	Technology	Science
Introductory Calculus (3-4)	x	x	x
Multivariable Calculus (3-4)	–	x	x
Probability (3)	x	x	x
Linear Algebra (3-4)	–	–	x
Theory and Techniques of Calculus (3-4)	–	–	x
Statistical Inference (3)	–	x	x
Introduction to Computing (3-4)	x	x	–
Principles of Programming (1)	–	–	x
Total hours	9–11	15–18	19–22

The Committee recognized the difficulties to be faced in reaching these goals, especially in view of the fact that it is rare to find more than nine semester-credits of college mathematics required for the bachelor's degree in agriculture and natural resources, and that it is common to find six or less. The following obstacles must be surmounted:

- Many faculty members do not yet appreciate the value of mathematics.
- The curriculum is already crowded.
- Many students have weak high school training in mathematics.
- Mathematics is not used as effectively as it could be in most substantive and supporting courses.

It was suggested that colleges of agriculture and natural resources stipulate that by 1972 students entering four-year curricula must have taken four years of college-preparatory mathematics.

The Committee justified its extensive recommended requirement on the fact that during the twentieth century probability theory, statistics, linear programming, and other branches of mathematics that have many applications in the biological, management, and social sciences have developed rapidly. The Committee report stated:

Probability theory provides mathematical models for the study of events with chance outcomes. It is the mathematical foundation for genetics and also for statistical theory. Statistics yields methods to summarize large collections of data and to draw conclusions from observations about events with chance outcome. Many important developments in statistics have been stimulated by problems in agriculture and natural resources. Linear programming provides techniques for solving many optimization problems which cannot be solved in any other way; for example, the formulation of a feed mixture which has specified nutritional values and minimum cost, the allocation of farm resources to maximize profits, or the allocation of natural resources to maximize public benefit. Linear programming in turn rests on the basic concepts of vector spaces and linear algebra. Some optimization problems can be solved, however, only by the older techniques of differential calculus.

The Committee pointed out that an increased mathematics requirement would actually improve the efficiency of undergraduate training by eliminating the repeated need for including instruction in techniques of mathematics, statistics, and computing in upper-division specialty courses. They estimated that "the contact hours to cover the desired topics in many areas of upper-division instruction can be reduced by at least fifty percent," if the students were adequately prepared before-hand.

The Committee report itself gives a topic outline for each of the recommended courses and provides an interesting section in which direct and indirect uses of the various mathematical functions in agriculture and natural resources are listed in detail.

SUMMARY

The recommendations of the committees, in the aggregate, include a minimum of 18 credits of chemistry, 8 credits of physics, and 9 to 22 credits of mathematics (depending on the curriculum)—a total of 35 to 48 semester credits. This represents a 65 to 140 percent increase over the average number of credits now taken by students in agricultural science and agricultural technology.

One can anticipate that recommended increases of this magnitude will be met initially with strong resistance from the administrators and faculty. However, it is my plea that we do not dismiss the entire project as "unrealistic." There is much here of great value.

I am favorably impressed with the attitude and spirit of cooperation among the people from the basic sciences. As individuals and professionals, they are dedicating their time and resources to the improvement of undergraduate teaching in the basic sciences. Many recognize the value that would result from the use of examples and illustrations that clearly have relevance in agriculture, natural resources, biology, and everyday life. However, they need help in securing source materials, appropriate examples and orientation to the developments that are taking place. Recommended course content must be brought to the attention of the responsible people in the chemistry, physics, and math departments.

F. YATES BORDEN

Under the auspices of the Commission three committees were consti-
tuted to evaluate the present situation, trends and future needs of
physics, chemistry and mathematics education in conjunction with
the anticipated future educational requirements in baccalaureate programs in agriculture and natural resources. They were to make recommendations concerning courses, course content and orientation,
instructional methods and materials, implementation and other related matters.

AREAS OF GENERAL AGREEMENT

Each of the reports stressed the rapidly changing educational environment of agriculture and natural resources and of physics, chemistry,
and mathematics and the need for frequent review to meet changing
needs. After all, an agriculturist can validly contemplate mathematics
education only by being aware of the concurrent changes in mathematics education. For this reason one can expect effective results
only if the group is composed of up-to-date representatives of the
various disciplines. That educators with such diverse backgrounds as
those participating in the conference could reach agreement on many
points lends support to a belief that common goals in collegiate education do exist.

Major areas of agreement follow:

• There is an increasing demand to quantify courses, curricula and
disciplines that are related to science and technology.
• Special courses with appropriate orientation and topic coverage
should be offered for students in agriculture and natural resources,
but these can also be considered appropriate for students in other
biological or life science curricula.
• Courses given for agriculture and natural resources students
should not be terminal, nor diluted, nonrigorous analogs of basic
courses in mathematics, physics or chemistry.
• Materials pertinent to such courses are at present not adequate
and measures should be taken to rectify this situation as rapidly as
possible.

- Communications between the agriculture faculty and faculties in physics, chemistry and mathematics should be strengthened.
- Continuing education in physics, chemistry and mathematics for agriculture faculty is a problem for which many solutions exist, but which can in fact be solved only when substantial emphasis is placed on faculty participation.
- Although secondary school programs are being strengthened in the basic sciences and mathematics, the qualifications of students entering agriculture and natural resources curricula are likely to be more diverse than in the past.

With regard to special courses it was generally agreed that courses for majors in chemistry, physics and mathematics would not often be appropriate. Service courses for nonmajors could and should be designed for students in biological and life science areas, including those in agriculture and natural resources. Such courses should not be terminal, so that students might continue with the intermediate and advanced courses without serious penalty. They would differ from the courses for majors mainly in orientation, supporting materials and frame of reference. Topic coverage and prerequisites would be substantially the same.

Continuing education of agriculture faculty was regarded of extreme importance in order to keep them up to date on changing patterns in mathematics, physics and chemistry courses and to strengthen their capacity to use mathematics, chemistry and physics material freely in their own courses. Continuing education of physical sciences faculty in the disciplines their courses support was also deemed important as, of course, was fluent communication between the two faculty groups.

PHYSICS

The Committee on Physics agreed that a one-year course in physics was adequate for students in agriculture. The course should be one specifically designed for them and would apply equally well for students in other biological sciences. The course, although distinct from a course for physics majors, must not be terminal nor superficial. It must be based on a working knowledge of calculus and preferably a background, not necessarily comprehensive, in modern algebra, probability theory and computer science. The main differences would be in the orientation and supporting materials. More emphasis would be placed on fluid mechanics, thermodynamics and elementary radiation than is usually true of a first course. Bonafide examples pertinent to the biological sciences should be used, although such material has not yet been assembled. The course in physics should develop in the student the "conviction that quantitative thinking in terms of a small group of widely applicable theoretical generalizations is a technique that is widely applicable."

As for concrete action, it has been proposed that pertinent source materials be developed as rapidly as possible and instruction modules be prepared to facilitate implementing the various recommendations.

CHEMISTRY

The Committee defined a full-year course of 10 semester hours in chemistry as minimum. The course would encompass the fundamental aspects of inorganic, organic, physical and analytical chemistry, but it should be neither a survey course nor a terminal course. A second course already outlined in the literature* was considered to be desirable in most curricula, but chief emphasis was placed on defining the first course and outlining the general topic areas of a two-course sequence. The first course lectures center upon four areas: (1) atomic theory, bonding, nature of molecules, gases, solids and liquids, (2) organic compounds, (3) chemical energetics, and (4) descriptive inorganic chemistry. The practical applications would cover a broad spectrum of topics from the areas designated earlier and should be strongly experimental in emphasis.

A student who took the two chemistry courses recommended by the Committee would have the equivalent of a one-semester course each in general introductory chemistry, organic chemistry, and quantitative analysis. In view of the physical chemistry and biochemistry included in the recommended courses, the student would also be well-armed to persue intermediate and advanced courses in any of these areas.

The Committee considered some of the problems facing present and future teaching of chemistry as related to agriculture students, faculty and curricula. Generally, their recommendations were very

*Everhardt, W. H. 1967. Newsletter of advisory council on college chemistry. May. Dept. of Chemistry, Stanford University, California. 94304

similar to those made by the physics group, but they did emphasize the need for deans and faculty of colleges of agriculture to recognize the need for continuing education of faculty. Numerous ways to implement continuing education were considered.

MATHEMATICS

The recommendations in the report by the Committee on the Undergraduate Program in Mathematics (CUPM)[†] were critically evaluated by the Commission's Committee on Mathematics. The courses defined by CUPM were designed as a basic undergraduate program in mathematics, providing in addition a general service to other departments. For a number of reasons, given in detail in their report, the Committee on Mathematics strongly advised the adoption of a number of the CUPM courses as requirements in the four-year B.S. curricula in colleges of agriculture and natural resources.

The agriculture curricula are categorically designated as education, technology and science—business programs are considered to be appropriately placed in either the technology or science curricula. All programs would require a course in introductory calculus and in probability. Other courses specified for one or more of the three categories include multivariable calculus, linear algebra, theory and techniques of calculus and statistical inference. Two courses in computer science were indicated by the Committee to complete the selection. (See table opposite.)

Although the semester-hour requirements appear to be very substantial, many specific curricula already installed in colleges of agriculture and natural resources have requirements nearly as demanding. It should also be pointed out that a number of these courses are often not considered part of the formal mathematics requirements. In the traditional classification of mathematics vs. nonmathematics courses, the education category requires only 3–4 semester hours of mathematics; technology, only 6–8; and science, only 12–16.

The Committee emphasized the need for mathematical ability at the level indicated for the bachelor of science 10–15 years. Judged by present standards in secondary school and in agriculture curricula,

---

†Committee on the Undergraduate Program in Mathematics. A general curriculum in mathematics for colleges. A report to the mathematical association of America. 1965. CUPM, P.O. Box 1024, Berkeley, Calif.

| Course Name | Recommended for Curricula in | | |
(Semester Hours)	Education	Technology	Science
Introductory Calculus (3-4)	x	x	x
Multivariable Calculus (3-4)	–	x	x
Probability (3)	x	x	x
Linear Algebra (3-4)	–	–	x
Theory and Techniques of Calculus (3-4)	–	–	x
Statistical Inference (3)	–	x	x
Introduction to Computing (3-4)	x	x	–
Principles of Programming (1)	–	–	x
Total Hours	9–11	15–18	19–22

this program can be installed by 1980 only if implementation were to begin immediately. The Committee confirmed the need for mathematics at a reasonably sophisticated level in the subject areas characteristic of agriculture and natural resources, but pointed out that undergraduate courses seldom make full or efficient use of mathematics, even when they purport to be quantitatively oriented.

One of the major obstacles to implementation of the recommended mathematics is the faculty of agriculture colleges. A number of these faculty members do not understand the applicability of mathematics to their areas of interest, a value that can be recognized only if emphasis is placed on continuing education of faculty and in a general and substantial increase in mathematics training employed in the graduate programs.

GENERAL PHYSICAL SCIENCES

Two working groups, one on integrated physical science courses and one on physical science and mathematics, were also formed. The latter recommended that no new special courses be required of every student in agriculture and agreed in principle with the mathematics, chemistry and physics committees' reports, suggesting that the chemistry and physics courses be specified in greater detail at some later time.

The working group on integrated physical science courses in essence strongly advised against the general use of integrated physical sciences course as substitutes for specific courses in mathematics and sciences. Two reasons cited in support were the actual failure of such courses to meet their objectives and that, in practice, no time savings have been realized. The group did, however, recommend that continued effort be expended in devising and implementing such a course, at least experimentally.

GEORGE A. GRIES

Everyone gripes—agricultural and natural resource educators are no exception. One of the favorite targets of their complaint is the nature of the courses in chemistry, physics, and mathematics available to their students. The old courses designed especially for "aggies" were watered down and too often assigned to the poorest instructor in the department. On the other hand, the standard courses for majors, while sufficiently rigorous, lack relevance and hence motivation for the agricultural students. Examples usually are drawn from theoretical aspects of the discipline or from engineering. Agricultural or biological illustrations are seldom, if ever, used—primarily because of the biological illiteracy of the professor.

The ability of the instructor in a physical science or mathematics course to relate his discipline to the interests of the student of applied biology would in no way reduce the responsibility of the professor of agriculture or natural resources to convince the student of the essential role of the physical sciences and mathematics in his education. This can best be done by demonstration during his own instruction; if he fails to take advantage of every situation that allows this, he has failed in his responsibility.

Dissatisfaction with the present offerings in mathematics and the physical sciences was forcefully presented in the reports of Commission's committees on the basic science training needs of students in the several agricultural disciplines in 1966. This was one reason why the Commission established separate panels in chemistry, physics, and mathematics in 1967. These panels were charged to make recommendations as to training needs in these basic disciplines for students who would be leaders in the agricultural professions ten to fifteen years in the future. They were invited to consider subject matter, teaching procedures, sequencing of topics or courses or any other aspect of the educational process they wished. Specific suggestions on how their recommendations could be implemented were requested. They were not asked to consider the curriculum as a whole.

All of the committee reports recognize that curricula in agriculture and natural resources have made significant changes during the last decade in an effort more nearly to reflect the needs of the industries. Most marked of changes is an increased emphasis on basic principles at the expense of skills, which have a tendency to become obsolete. The need for greater sophistication in the future, even for the terminal bachelor's programs, is noted, dictated by the increased application of modern technology. The reports also point out that rapid increases in knowledge has necessitated a change in instructional programs and that, since agricultural and natural resource curricula must be based on sound modern concepts and principles, educators in the applied sciences will be forced to keep abreast of these changes. The development of strong nonterminal courses in the basic sciences, specifically designed for those in the biological disciplines, depends primarily on the availability of teachers with insight and interest. The need for source materials on the biological application of basic mathematical, physical and chemical concepts was recognized in each of the reports.

The Chemistry Panel suggested a minimum of a one-year course (10 credit hours) for students in all agricultural and natural resource curricula, regardless of their educational goals. This course should have a strong experimental orientation, should clearly show the impact of modern chemistry on society, and should present an adequate overview of the whole field of chemistry. The early introduction of concepts of organic and biochemistry would assist in making the course more relevant and give an essential background for students not electing further courses. It should not be a terminal offering and should presume a high school preparation of at least three years of mathematics. Suggestions for laboratory topics were given. The need for a sourcebook or other resource for chemistry teachers without adequate background in biological, agricultural, or natural resource disciplines is emphasized. The Panel suggested that the second course in chemistry should be similar in content to the course in "bioganalytical" chemistry currently being offered at U.C.L.A. Students in science options should be able to proceed directly from this course to advanced offerings in chemistry.

The report of the Physics Panel shows but little enthusiasm for present introductory courses in physics. The members suggested that it should be possible to provide a substantial course that is both practical and sympathetically presented without catering to the interests of the physics major or the engineers. It should be of such depth and based on sufficient competence in mathematics to allow students to move into conventional courses at no great cost in academic effort. Such a course would appeal not only to students in such applied biological disciplines as agriculture, natural resources, and the health-related fields but also to those in basic biology and in the liberal arts. The present dearth of competent physicists who are literate in biology could be overcome in part by the preparation of source materials upon which the teacher could draw. The Panel recommends that at least two pilot projects be initiated to develop teaching modules and to try out courses similar to the one they envision.

The Panel on Mathematics began its report with an analysis of the needs of the agricultural and natural resources student of ten to fifteen years hence, as he prepares himself for the industry as it will be during much of his career. The panelists visualized that students in agricultural and resource education, technology (production, management, and business) and science will all need greater sophistication than they now have. The members of the panel suggested that mathematical courses similar to those described in the General Curriculum Report of the Commission on the Undergraduate Program in Mathematics would be entirely satisfactory units from which meaningful sequences could be built. They suggested that minimal mathematical training for all students in agriculture and natural resources should include introductory calculus, probability, and a course in programming or computing. Students in the technological curricula should, additionally, have work in multivariable calculus and statistical inference, and students in science options should have courses in those areas and in linear algebra and theory and techniques of calculus. Total semester credit-hours range from 9-11 to 19-22 in the different options. The Panel recognized that there will be substantial resistance to the incorporation of so much additional mathematics into an already crowded schedule but believed that there would be an equivalent saving of time and effort presenting technical subject matter to students with stronger mathematical background. This, of course, implies relatively high mathematical literacy among those who teach agriculture and natural resources courses.

The Panel recommends the development of resource materials and suggests ways and means of both developing them and of providing needed training to mathematics and applied biology faculty members.

5

Social Sciences

CARROLL V. HESS

American communities are today faced with massive social problems. The annual level of investment in the development of technology has reached the 23 billion dollar level in the United States. Seven billion of this is investment by the private sector; the remaining 16 billion represents public investment in technology. Investments of this magnitude generate enormous changes in capital input supplies and engender growth in goods and services of all kinds. As firms operating in this capitalistic society respond to huge injections of technology, there are increased specialization, economies to scale, and marked structural changes within the economy.

Many communities are unprepared to meet these massive changes. As one measure of public concern, witness the fact that in 1946 the federal government spent 894 million dollars to help local governments augment their public programs. By 1966, this figure reached over 14 billion dollars, a 16-fold increase in two decades.

A myriad of community needs have to be faced—equitable taxes; educational expansion and relevancy; adequate local government;

85

racial reconciliation; elimination of poverty; modernization of laws
and public codes; improving the quality of the environment; maintenance of public order under newer structural and social systems and
conditions; adequate health services; better transportation and communication; and, finally, the human desire for beauty, dignity, and
well-being. Adequate progress toward these goals can come only from
citizens whose understanding is in perspective with the times. Science
and education must be as uniquely organized for the purposes of
social innovation as they had to be for technological innovation.

Our land-grant universities were developed during a period in history that demanded expansion and application of the biological and
physical sciences. The agricultural and forestry colleges, with their
experiment stations and cooperative extension services, were among
the great innovations in American social and economic history. This
system of research and education through the invention and introduction of new technology was an important instrument in our developing nation. The very technologies that freed us—or many of us—
from hunger and want has helped create major social issues and conflicts. Thus we must consciously seek the proper place in agricultural
and natural resources curricula for the social disciplines to contribute
to an understanding and of the social issues arising as a kind of technological backlash. Technology has thus been a mixed blessing, by
no means as yet subjected to effective social control. Furthermore,
there have been few if any periods in which our social structures have
been more critically challenged for their internal inconsistencies and
inequities. Thus, the social sciences are directly charged with the
study of a society that is today in turmoil. Social sciences, as used
here, refers to economics, sociology, anthropology, political science,
geography, and psychology.

The social sciences focus on man's behavior and the ways in which
he interacts with his fellowmen through social, economic and political
institutions to implement his aspirations and values. Thus it is to the
social sciences that man must turn if he is to assert his dominance
over his technology and face constructively the manifold social issues
of our time. Without reasoned and objective approaches to these
matters, we can rely only on expedience and luck to avoid chaos.

The Committee recognized that agricultural and forestry colleges
are properly concerned with professional education. At the same
time, the student must live his professional and personal life in an ex-
plosive world with its unresolved issues. While education in the social
sciences cannot always yield immediately applicable social knowledge,

it is vital to alert students to the man-created conditions that will persistently impinge upon the practice of their professions and their lives in the community.

Notable recent improvements in secondary school education have occurred in mathematics and biological and physical sciences, developments that have not been matched in the social sciences and humanities. Thus colleges and universities have accordingly been faced with the increased burden of sensitizing students in these disciplines to their changing social milieu and assisting them in interpreting it.

Although modern society has profited from the advances that have been made in the physical and biological sciences, it has become increasingly difficult to adjust old institutions or to establish new ones that take full advantage of the new technology. The pressures created by our burgeoning populations have placed special emphasis on land-use conflicts and on the need for increased understanding of the complex socioeconomic culture in which we live. Modern facilities and students alike are concerned about an educational process that produces graduates filled to the mortar borad with "scientific" facts, but who fail to understand the nature of man and of the institutions man creates in his attempt to optimize his individual and societal goals.

There is growing awareness that an agricultural and forestry "professional" is more than a "technologist." It is widely recognized that he must be an educated man who can adapt rapidly to changes in both technology and social goals and that to meet this challenge he needs a more liberal type of education than has traditionally characterized agricultural and forestry education programs. One response to this need is the growth of two-year programs that produce technicians who free the professional to concentrate on decision-making, policy formulation, and public planning.

The variety of subject matter and career opportunity in agriculture and natural resources encourages some flexibility in developing social science training for agricultural students. The social science contributions to the production-technology or business options, for example, may be very much applied in nature, serving an "instrumentation" function. On the other hand, it may be wiser to use social science courses to create a philosophical background that will help students appreciate both the biological science teachings and the application of them in the service of humanity.

Likewise, in the natural resources area there are special needs.

"People problems" involving political action, aesthetics, use preferences, and qualitative judgment have replaced the old questions of
sustained yield, habitat, manipulation, and methods of inventory.
This transition is especially evident in outdoor recreation, a newly-
emerging field of resource management. As a result, educators in resource management are calling for greater social science input to
their undergraduate programs.

In short, today's graduate of agricultural and natural resources
management programs works largely with complex problems requiring a broad interdisciplinary background for their solution. He
must be able to communicate with, and understand, something of
the methods of engineers, planners, and social scientists who contribute to the analysis and solution of his problem. The undergraduate in
agriculture or natural resources must have increased exposure to key
subject areas in social science if he is to help these teams of specialists
achieve their full potential.

RECOMMENDATIONS

The Committee emphasized the need for flexibility that would allow
a student to add new disciplines, including the social sciences, in line
with the increasing maturity of these students in professional colleges.
Social science courses taken early in the bachelor's program expose
the student to a new fund of knowledge as to how social scientists
perceive the world, as well as sensitizing him to a wider range of vo-
cations and life styles. Flexibility also permits the student to fashion
a program compatible with his specific professional needs and goals.

The Committee specifically recommended that undergraduates in
agriculture and natural resources devote 15 to 20 percent of the curriculum to the social sciences, that the courses extend in depth in at
least two of the six social science disciplines, and that the general
education experience in the social sciences be similar for all professional specializations.

In most instances, the recommended 15-20 percent allocation to
the social sciences represents a significant increase, but one that is
justified because of the greater progress made at the high school level
in the biological and physical sciences than in the social sciences.
Little hope exists there for any change in emphasis on the social sciences, thus leaving it up to the colleges to meet the social science
needs.

The Committee felt that today's baccalaureate candidates in professional schools should emphasize more social science if they are to be prepared for the world in which they must practice. The inevitable reduction in professional courses that occurs seems justified on grounds that the present level of professional over-specialization at the undergraduate level has been won at the cost of sacrificing a solid general education, basic-science exposure.

The Committee argued against excessive open electives because of the tendency for students to utilize those within their major. Thus, a broadened program that encourages use of previous free elective credits outside the major seemed essential; still a maximum of 30–36 percent of total credits in the major was recommended.

The Committee went still further and suggested specific courses in applied social sciences for certain "vocational or job-oriented" programs—for production-technology options, for business and industry options, for education-public administration, for graduate school preparation. Suggested courses were group dynamics, policy formulation, political and economic development, business law, personnel management, developmental psychology, social and cultural change. The report also includes a descriptive summary of how each of the six social science disciplines relates to and can contribute to the undergraduate education of students in agriculture and natural resources.

Finally, the Committee encouraged communication among faculty members in agriculture and natural resources with their colleagues in the social sciences and throughout the university to effect the best possible combination of existing, restructured or new courses in the social sciences.

6

The Junior College

ERNEST TARONE

There seems to be a very healthy interest in the junior college vis-à-vis its relationship to and meaning for agricultural education. I and other junior college people have great concern for the relationship of our curriculum with the four-year institutions. Most junior college instructors are working under less than ideal conditions in trying to relate to the four-year institutions, and junior college people very much want cooperation from the four-year colleges in order that the best possible job may be done in the interest of the students. There is a need for the four-year institutions to recognize the junior college, its philosophy, the job it is trying to do, and how it relates to the four-year programs.

The junior college has diverse functions—the transfer function, the terminal (occupational) function, and the counseling function, the last being considerably better developed than in most other types of institutions because of the variety of students it tries to serve and the needs that they have.

There is much misunderstanding of the junior college movement, which leads to a number of "myths." This is evident in the names sometimes given to the junior college—"high school with ash trays," "glorified high schools," the "democratic educational institution," the "open door college," the "terminal college," the "creative college," the "commuter's college," and the "citizen's college." All of these terms conjure up a very mixed image of what is, in fact, the only truly American educational institution and, at the same time, the fastest growing educational institution in the United States. The great variation in kinds and types of junior colleges and in their methods of support and control fuels the mythology but has also an element of fact, varying with the institutions. More undergraduates in California (approximately 66 percent) are in the junior colleges than are in the two-year lower division of state colleges and universities; this percentage will increase. In the West particularly, states are adding junior colleges at an increasing rate, and in such states as California, Illinois, Florida and New York we have reached 85 percent of the goal of having a junior college within commuting distance of every student. By 1975 many states will have achieved this goal.

The junior college is unique as an educational institution. Its one over-riding function is instruction—not research, not writing, not publishing. While in most states scholastic achievements determine who may attend a state college or a university, the junior college attempts to handle all grade levels and meet all needs. As junior colleges grow, four-year colleges will be getting more students as juniors than as freshmen, students who have been prepared in junior college. It, therefore, behooves the state colleges and universities to try to comprehend the junior college, its staff, and its students.

Ten items, imaginary or otherwise, have been suggested to me by four-year college instructors as typical of what they think of as characteristics of junior colleges. My response follows in each case.

- *Junior colleges duplicate four-year college teaching.* But the teaching duplicated is not necessarily at the same level or with the same objective. Eventually a great many of the students are fed to four-year institutions, having gained in maturity, learned how to study, and found that they can achieve or that they have ability.
- *Not all students have the prescribed prerequisites.* But we transfer students, not courses. In California mechanisms have been worked out that lead to transfer of students with the same kind of background and who have the capacity for given levels of work, rather than primarily those who meet certain prerequisites.
- *The junior college tries untested techniques.* Certainly, new things are tried out by the junior colleges—new directions in the

application of natural resources principles to recreational land management, or field application of long established principles of natural resources. The junior colleges, unhampered by tradition, are less inhibited.

- *Those teaching agriculture seldom have background comparable to four-year college teachers.* True enough, teachers who have not had animal nutrition do teach feeds and feeding. But why cannot one teach basic principles better in a practical situation than in an unreal one? It seems entirely possible that a student might learn the facts about nutrition best when he actually feeds animals and develops rations.

- *The students should emerge from junior colleges with all necessary math, chemistry, English, biology, etc.* But when they get to the four-year school, they are still in agriculture, not in biology or pharmacy or some other field. If they are to remain interested in it then, they must get some agriculture, lest they get diverted into "more worthwhile" things.

- *If a school has 40 courses in agriculture and only 3 teachers, its courses do not deserve to be considered except as electives.* One must remember that it takes years to build a program. The job of building significant programs at the junior college level is extremely difficult and needs help from all sides.

- *Junior college counselors are told by administrators of four-year institutions simply to "follow our catalog."* But what about the two thirds of the students who do not go on to the four-year college? What is really needed is articulation paths that will help get the student to the four-year institution once he is ready, not blind adherence to a catalog.

- *Why do not potential transfer students take the "right" courses?* How can the junior college instructor or counselor know which ones will transfer? Many of those who say they will do not; many of those who say they will not end up doing so. Sometimes they remain in the junior college two years before reaching a decision.

- *Junior colleges are placing students in jobs that would otherwise be available to four-year graduates.* We do try to train students directly for identifiable need in the community, hopefully with a curriculum that fits the job to be done. Placements from junior college are not done with the express intent of depriving graduate students of jobs but to meet a need that industry has.

- *Job titles do not fit.* We use what we find in the market place, whatever has meaning in relation to the work to be done. We have to be understood in the community; we are close to the taxpayer.

J. CLYDE DRIGGERS

This report refers specifically to articulation between Abraham Baldwin Agricultural College, a two-year college oriented toward agriculture, and The University of Georgia. Although each institution has a high degree of autonomy, budget policies are determined by the Board of Regents of the University System of Georgia. Articulation has succeeded mainly because there is close cooperation and understanding between the officials and representatives of the two institutions. Without it, no articulation guidelines, regardless of how well conceived and presented, will assure smooth and satisfactory student transfer.

There has been adopted within the University System of Georgia a so-called "core curriculum" for the first two years of work that is applicable to all college parallel programs, regardless of whether they are taken at a junior college, senior college, or university. Students who graduate under the core curriculum are accepted without loss of credit at four-year colleges and universities of the system. They normally will be accepted also by all other fully accredited colleges and universities of the nation.

The areas of study and the minimum quarter hours required for each are as follows:

I.	Humanities, including but not limited to grammar, composition, and literature	20
II.	Mathematics and the natural sciences, including but not limited to mathematics, and a 10-hour sequence of laboratory courses in the biological or physical sciences (may be a laboratory in a behavioral science)	20
III.	Social sciences, including but not limited to history and American government	20
IV.	Courses appropriate to the major field of the individual student	30
	Total	90

In the area of general agriculture, the 30 quarter hours of electives may be selected from the following agricultural offerings: Agricultural Economics, Crop Production, Animal Husbandry, Soil Management,

Engineering Graphics, Drainage, Irrigation and Erosion Control, Farm
Power and Equipment, Farm Electrification, Surveying, Farm Forestry,
and Poultry Production.

Within the junior colleges, where no agricultural courses per se are
offered, the 30 hours of electives in agriculture may be satisfied by
selections from the following: zoology, botany, biology, chemistry,
economics, horticulture, mathematics, physics, and statistics.

To satisfy the 30 hours of electives in agricultural engineering,
forestry, and home economics, the student majors must take those
courses especially designed for and required of them.

In states where there is no university system or core curriculum
per se, it is recommended that articulation be initiated between the
appropriate officials and representatives of the two-year and four-
year institutions by adopting curriculum requirements that are offered and can be accepted by each institution. When this is done effectively, there will be few dificulties.

7

Industry's View

H. L. WILCKE

The subjects before this conference are vital to young people interested in agriculture and who intend to make agribusiness their life's work. They are also vital to many who are not directly concerned with agribusiness but whose work will require an understanding of just what it is, what it encompasses, and what its purposes and goals may be.

Just a few years ago Webster's 3rd defined agriculture as follows:

The science or art of the production of plants and animals useful to man and in varying degrees, the preparation of these products for man's use and their disposal (as by marketing).

Compare with this the following statement by Dean Sherwood O. Berg of the University of Minnesota:*

*Institutional Research Service. 1969. Farm outlook report, June 26. Piper, Jaffray, & Hopwood, Minneapolis, Minn. 3 p.

The scientific, technological and knowledge revolution which has been reshaping the face of all American industry has resulted in great benefits in the efficiency of mass production of food and fiber. The major benefits have accrued to the American consumer—that of a plentiful, relatively low-cost, highly-varied food and fiber supply. While the major benefits of this fast-moving revolution in agribusiness are recognized and generally appreciated, another aspect of the revolution is far less well understood and appreciated. This is reshaping the *structure* of the agribusiness industry.

The latter changes have been all-encompassing in the agribusiness sector—from the industries and firms supplying vital production inputs to the numerous industries and firms supplying vital production inputs to the numerous industries and firms which receive, transform and distribute the food and fibers. To understand the long-range prospects for the agribusiness "system," one must look at the changes which have taken place and the underlying forces giving rise to these developments.

The agribusiness system is characterized by a series of highly interdependent and interrelated stages and processes, where value is added as one moves from the initial input stage to the final consumer. Recent estimates indicate that, of the $100 billion spent annually for food, 33% was represented by supplier sales to farm producers, 12% value was added in farm production, and 55% was added in the food processing and marketing sector.

Growth in the agribusiness system has been most spectacular. Since 1950 the retail value of sales of all U.S. produced food has doubled from $50 billion to $100 billion. In the same period, the purchased inputs (petroleum, feed, fertilizers, seed, insecticides, etc.) have increased by over 35% and employment in the food sector has risen by 15%. The food and fiber processing sector has experienced a 25% increase in employment since 1947.

Several other major trends and developments are also apparent:

1. *Concentration of firms.* The farm-supply and marketing sectors have experienced a reduction in the number of firms but an increase in firm size. An example is the 6% decrease in firms manufacturing food products in a recent 15 year period, while firms with 100 or more employees have increased by 25%.

2. *Vertical integration* is well established among such commodities as broilers and fruits and vegetables and is increasing in beef cattle feeding and in egg and turkey production.

3. *Conglomerate integration* is the acquisition and control of other economic enterprises not necessarily closely related to that of the integrator. There are several illustrations of food processing firms acquiring firms in the industrial manufacturing, aerospace and transportation and vice versa.

4. *Shift from product emphasis to systems emphasis.* An example of this shift involves firms which two decades ago may have visualized their operations as processors and sellers solely of dairy products but who have since broadened their economic functions to encompass many processed food products. This may have occurred even if traditional dairy product lines were dropped. The systems emphasis has hastened the adoption of automated processes to coordinate firms' activities. The functions of these companies have, moreover, been modified to maximize money flow and profits through contracts and expanded ownership.

These major changes in the U.S. food and fiber system have given rise to agri-business firms providing a wide range of products and services with stepped-up efficience. Value added per employee, a rough indication of efficiency, has increased 154% since 1947 in the farm input sector and 261% in the product sector. Capital expenditures have increased 90% in the input sector and 164% in the product sector during the same period. The added emphasis on service and new products has been reflected by a 143% increase in profits in the input sector and a 66% increase in profits in the product sector.

The events of the last 20 years reflect the ever-changing nature of the agri-business sector of the U.S. economy. Clearly, the increasing scientific base, the developing specialization and greater capital investments have a significant relationship to the market demand for food and fiber.

This description of agriculture by Dean Berg provides a fine concept of just what agriculture is today and it underscores a new dimension—those who supply the man who is producing the plants and animals. It also emphasizes the food processing aspect, which is a part of agribusiness. It emphasizes that agribusiness has become a business rather than an art. It provides a much better basis for understanding just what we are preparing for—the kind of training we need.

For the biological sciences, turn again to Webster:

The science of life, a branch of knowledge that deals with living organisms and vital processes broadly including zoology, botany, morphology, genetics, embryology, and allied sciences but commonly being restricted to consideration of principles of wide application to the origin, development, structure, functions and distribution of living matter as represented by plants and animals and to the generally recurrent phenomena of life, growth and reproduction.

There are those, particularly in biology, who feel that agriculture is simply a part of biology. I reject this concept totally. I think we may regard agriculture as a system that employs not only biology, genetics, morphology and so forth, but also chemistry, physics, mathematics, and, in fact, most of the basic sciences at some point in agribusiness. It seems to me that in this regard agriculture may be compared to the fields of medicine or engineering, each of which cuts across lines of all of our basic sciences, utilizes them, and applies them to the problems of producing food and fiber.

There are those who feel that discoveries in the basic sciences require an intellectual brilliance that is not necessary to the application of knowledge, but was not Eli Whitney fully as brilliant in applying principles to develop the cotton gin as those who discovered those principles in the first place? Have not our space people been just as brilliant in applying the principles of physics, mathematics, engineering, and astronomy to the task of landing men on the moon? Is it not one of our jobs today to convince young people that there is a challenge, that there are opportunities in agribusiness, and that they must prepare themselves just as thoroughly and intensively as for any other line of work?

Perhaps the best way to convey the industry attitude toward the teaching of the basic sciences to students in agribusiness would be to give you some of our impressions of today's graduates and to indicate some of the ways these young men and women develop in a business organization.

In recent years every facet of agriculture has had to develop a strong business approach to its operations. The application of business principles is critical not only to the marketing of end-products but to production, manufacturing, distribution—even research and development.

Our universities are making good progress in supplementing the student's agricultural background and education with a basic understanding of business principles. We wholeheartedly endorse these efforts and would encourage any possible acceleration.

Most students today are more mature, knowledgeable and perceptive than at any previous time. With proper guidance they can be more flexible and responsive under a wide variety of conditions. Young men have asked for the opportunity to get involved sooner, to share responsibility, to do things. And they are doing a surprising job of shouldering these duties. Because of this, lengthy training programs throughout industry are being sliced. New programs are being put into action that will provide a brief exposure, plus on-the-job development.

At the undergraduate, or even the master's level, American industry is not so deeply concerned with a young man's specific skills as they are with his interests, response, drive, ability to communicate and desire to become personally involved. We prefer that he have an agricultural background and formal education, but to perform most effectively, these other attributes are vital.

Food companies are particularly interested in a man's industriousness, his interest and desire. As you well know, the food industry has always been a low-margin industry, even more so today than in the past. Men will often carry multiple responsibilities; there is seldom a situation where a department or group or division is over-staffed. By modern business standards, this is wise and healthy. . .but it requires a high degree of performance and flexibility from each individual.

We talk to an alarming number of young men who are thoroughly confused about their future. They are considering business, but apparently only as a last resort, and businessmen do not want to employ persons with this attitude—a situation that produces a communications roadblock between student and recruiter. The man is staunch in his desire to work more directly in his chosen field; the industry representative is searching for a spark of flexibility, for a sincere interest in a business career.

We frequently see a direct correlation between student attitudes and faculty attitudes. Where faculties have developed a close working relationship with industry and have become an integral part of commercial agriculture, students have a greater awareness and interest in the agribusiness complex. To some extent we could criticize agribusiness for failing to shoulder its share of the communications load and help bridge this gap.

The need for men with strength in the basic areas is obvious. The current manpower market is tough and the willingness of agriculture-related industries to obtain men with these attributes is reflected in starting salaries. The food industry in general had a 9 percent starting salary increase two years ago (at the undergraduate level) compared to 6½ percent to 7 percent for other industries. The salary figures from the most recent recruiting year also show a comparable spread. In spite of the tight market, most all managements agree that they would rather leave a position open for an extended period than fill it with an individual who really does not fit. Not only are salaries rising, but the range is spreading, another indicator of selectivity.

At one time industry clearly had a decided advantage in attracting graduate students. More recently the contribution of government and educational institutions to American agriculture and the remainder of the world has been recognized, and salary differences and factors have been minimized. These salary adjustments reflect supply and demand—free enterprise at work.

Career opportunities are infinite, but they vary greatly, making flexibility and a broad background essential. The functions and responsibilities a man eventually shoulders can be very far removed from his initial role with a corporation. Even though tied to a company that is deeply involved in basic production, his duties might eventually encompass people, marketing, communications, money, materials, or any combination. Only a few starting positions offer the broad exposure that a company likes to give a man during these first few years.

What we have said, then, is that training for the profession of agribusiness must be broad, because agribusiness itself encompasses many different disciplines. We have said that the future leaders in agribusiness will be those who have been given the capacity to think and reason from a broad base. We have said that we are not as interested in specialization as we were a few years ago. Industry is interested in young men and young women who will challenge the old but who will not discard it entirely until they have found better ways and means of doing things. We are interested in those who challenge, but who challenge with reason and not simply with force.

The basic sciences are as important as they ever were, if not more so, in our agricultural curricula. They are important not as specialty entities but to provide a broader base, an understanding, of the problems and responsibilities that go along with providing the food and fiber for this country and others. Agriculture and agribusiness face a very real challenge in telling the story of what is really all about, in not being overly defensive when fadists and publicity seekers appeal to emotion in putting across their own particular viewpoints. Agribusiness needs leaders who are broadminded enough to recognize criticism and meet it, smart enough to foresee that there are many facets to a problem.

8

Summary Statements

P. J. LEYENDECKER

We should recognize that undergraduate training in agriculture and natural resources has a strong interdisciplinary orientation—that agriculture and natural resources are systems with complex subsystems that set interdisciplinary forces into action. Fortunately, we are flexible enough in our thinking to accept recommendations that can be implemented within the university environment in which we operate, recommendations on the minimal requirements for mathematics and the biological and physical sciences in the undergraduate curriculum. I believe this is the first time that such a group as this has attempted to bring together the thinking of those concerned with education in agriculture, natural resources, and the basic sciences. One might well ask, "Have we in agriculture and natural resources really made a realistic appraisal of our own course content in light of the curriculum changes which have and are now being made in biology, chemistry, mathematics, and physics?" With these thoughts in mind, I will review and summarize the highlights of the four committee reports.

CONTENT OF BIOLOGY IN THE CURRICULA

It was the general feeling of the two groups involved that there is no single approach to course content and that practicable solution would have to be hammered out between the biologists and agriculture and natural resource people at each institution. One group accepted the notion that an integrated general biology sequence should contain three major sections—cellular, organismal, and environmental biology—and that minimal requirements should be the same for all students of agriculture and natural resources—as an exception, the social science majors might be permitted to spin off at an earlier time in the minimal core requirements. One group was unwilling to allocate percentage times to course content, while the other group suggested percentage breakdowns and minimal core content requirements. No order of presentation within the minimal core requirement was suggested.

QUANTITY OF BIOLOGY IN THE CURRICULA

Both groups discussing this topic agreed that the minimal core in general biology should be a one-year course sequence for all majors in agriculture and natural resources. They did not feel that majors in social science or related areas should be permitted to bypass the minimal core. Minimal core requirements should be identical for all levels from B.S. to Ph.D. It was agreed that there should be flexibility in upper and lower division emphasis, depending upon the student's major, not a definite number of upper and lower division courses for all students in agriculture and natural resources. Strong student advisement was recommended to guide selection of upper and lower division courses in biology to fit specific major fields and the level of degree participation.

BIOLOGY AND THE TWO-YEAR AND FOUR-YEAR CURRICULA

Both groups assigned this topic agreed that the same general biology minimal core curriculum should apply in two-and four-year undergraduate programs. One group felt that there might well be three levels of biology course offerings, namely: professional, for the student going on into a four-year program; technical, for the student in the two-year program; and general biology, for the nonmajor. The other group was of the opinion that this could better be narrowed to professional and technical biology, and if limited to one core course, an arrangement should be made for a special tutorial laboratory slanted toward technical biology. It was emphasized by both groups that transfer problems could be solved by closer articulation between two- and four-year programs through seminars, workshops, and the like.

CHEMISTRY, PHYSICS, AND MATHEMATICS IN THE CURRICULA

The group was in general agreement with the Commission reports, namely that one year of chemistry should be the minimal requirement for agriculture and natural resource majors, and consideration given to the inclusion of some biological, quantitative, and organic chemistry in the one-year core. This course should be as vigorous as that required for chemistry majors. The group recognized a continuing and expanding requirement for mathematics up to and including calculus, with remedial work at the college level for students who need it. One year of physics was recommended, with certain modifications to meet more adequately the needs of agriculture and natural resource majors. This group also spent considerable time in discussing how to implement the recommendations. Suggested measures included discussions with recognized professional organizations and more interplay on home campuses through staff members in agriculture and natural resources taking or auditing courses in mathematics, physics, and chemistry. It was recognized that there is just no substitute for good teaching in all disciplines to bring about program implementation and curriculum change.

BURTON W. DeVEAU

QUESTION 1

What are the broad areas of biological knowledge on which agricultural and natural resources technology is dependent, and to which all students in agriculture and natural resources should be exposed? Which of these areas needs greatest emphasis?

All five discussion groups recommended that all students in agriculture and natural resources be required to take a one-year course in biological science. This course should include basic study in growth, reproduction, and behavior and a sequence of ecology, cellular biology, genetics, physiology, nutrition, morphology and evolution. Major emphasis should be centered upon the basic requirements of living organisms—reproduction, development, growth, and adjustment of behavior to environment. Appropriate sequence of courses is of importance if the interdisciplinary impact is to be maximized.

QUESTION 2

In what ways are the biology needs different for the various majors and options in the College of Agriculture—business, technology, and science?

Four of the five study groups recommended that all students in all options of agriculture and natural resources take the same first-year course in biological sciences. The fifth group recommended that students in such nonnatural science areas as home economics, agricultural engineering, and agricultural business should take a special terminal course in biology designed to their needs and including cellular, physiological, ecological, and genetic relationships in biology.

Several groups indicated that all college of agriculture majors should take additional courses oriented in biology, depending upon their needs and that those students majoring in applied areas of biology need courses in genetics and physiology.

All groups emphasized the need for appropriate guidance and advising of students. Several emphasized the need for frank, concerned dialogue between faculty in agriculture and the biological sciences to develop a fuller appreciation of the needs and desires of the two areas and to bridge the gap between the basic sciences and the applied sciences.

QUESTION 3

Should biology courses in the two-year technical program be designed so that they are transferable to baccalaureate program, or are the requirements of a technical curriculum such that the biology courses should be different from those in the bachelor's program? If they should be different—how?

The concensus of the study groups was that if a two-year program is terminable, its courses in biological sciences need not be designed to be transferable. Rather, they should be of an applied nature. If the two-year program is geared to educate students for transfer to a four-year program then the biology courses should be similar to those offered in the four-year institution. Transferability should be determined by the department granting the transfer credit.

One study group indicated that many two-year students have successfully transferred and have performed well in B.S. degree programs. They felt this attests to the caliber of the course material and presentation, and the varied application of the material in a way that makes the entire area meaningful and appropriate. This group further indicated that too much time may now be wasted with minutiae and worry over the content of the basic courses (and transfer credit), which seemingly do not hamper a student's progress if he is properly motivated.

QUESTION 4

How can the knowledge that the student gains from the various courses in the curriculum be integrated to best prepare him to perform professionally?

Integration of knowledge from various courses must be accomplished by each student on his own. To help in this task should be the objective of the faculty.

In order to enhance integration of the material in the basic courses all teachers must be well informed of the material covered, able to utilize it in their courses, and willing to eliminate unnecessary prerequisites to their courses. It is equally important that faculty teaching the basic courses be informed as to the uses being made of their course material in agriculture and natural resources. Thus, communication needs to be established between teachers of basic courses and teachers of advanced and applied courses.

It was essential that a strong faculty-student advising system be established and maintained; advising can play an effective role in the integration of courses and knowledge. Experience, through placement training, laboratory, and field contacts, helps the student to relate and interrelate his formal studies.

Students should be encouraged to think in breadth, to meditate, and to philosophize on the implications of their learning throughout

their academic career, not merely during the last term of the senior
year.

Several additional specific projects were recommended:

● Honors programs—Special topics for superior students involving
close faculty-student dialogue.

● Seminar classes—Utilization of team teaching and involving reports by students on specific interrelated topic areas that are modern,
relevant, and integrated.

● Open seminars—Campus-wide programs utilizing established experts in interdisciplinary topics.

● Summer field study tours—Special field tours that require the
student to observe and integrate the factors that impinge on a complex agricultural or natural resource region or situation.

● Formally organized interdepartmental courses.

● Greater utilization of student research projects.

● Intercollege integration—Breakdown of the artificial barriers
existing among colleges.

QUESTION 5

How can the recommendations of this discussion group be implemented?

It was generally agreed that discussion group members can be most
effective in implementing recommendations by sharing ideas with
their colleagues and by closely scrutinizing their own activities to
determine their relevancy and effectiveness. Active discussion and
dialogue among faculty with background and responsibilities similar
to those represented in the conference is essential. Successful implementation will require mutual objectives, understanding, and confidence among the different faculty and disciplines involved. Implementation may be enhanced by:

● The use of joint appointments between departments in core and
applied fields.

● The development of intra- and intermural dialogue, including
the exchange of course syllabi or outlines to best orient one another
to course content and offerings; adherence to re-evaluated prerequisites
for advanced courses.

● Broadening of agricultural courses to gain university-wide recognition, appeal, and acceptance.

• Establishment of all-day meetings at local institutions where faculty from agriculture, biology, and natural resources may discuss implementation of the basic biology core at the local level.

• Establishment of a system whereby faculty can participate in continuing education.

• Increased recognition of superior teaching and advising as key primary objectives of the total university program.

THOMAS W. DOWE

Broadly speaking the three basic approaches to the teaching of biological sciences are the evolutionary, the ecological or environmental, and the molecular. Much of the discussion here has been on what extent each of these should be stressed in a balanced biology curriculum; undoubtedly some of all should be included. But the general consensus seems to be that major emphasis should be place on ecological—environmental and molecular approaches. Attempting to strike a balance between these two is the problem that confronts departments and institutions. The emphasis on these two will depend on the character, the history, and the goals of the individual institutions.

I think it unwise for all institutions to come up with similar programs. There must be room for flexibility and differences. For example, in certain institutions ecological or environmental biology will be stressed far more than the molecular—institutions that emphasize the population explosion; the competition for land, air and water; alternative uses of natural resources. For reasons of their own, other institutions will tend to favor a more molecular type of biology. But even within these two broad concepts we find less than full agreement as to how the problems should be approached. For example, should we go from molecule, to cell, to organisms, to population; or from population, to organism, to cell, to molecule?

Another of the questions faced by the various groups had to do with introductory courses. We must first ask ourselves: For whom are these courses intended—majors or nonmajors? Obviously, at most institutions the majority of the students will be nonmajors, and I do not favor the organizing and structuring of introductory courses on a major—nonmajor basis. I feel that a good course cannot hurt or delay the major; I feel it exceedingly important for the nonmajor to receive the very best possible instruction in the basic biological sciences, regardless of his ultimate academic emphasis.

It is easy to say that we can structure courses for majors and non-majors and have them coequal. It is not nearly so easy to do it and have them remain coequal, because right away the majors begin to say, "The course for majors is better, more basic, and has in it a greater wealth of fundamental biology than the course for the majors." Besides, there may be a tendency to make courses easier when they are structured for nonmajors, so that the student can get a bit of this and a bit of that and go away thinking they have received a lot of something.

Let us not delude ourselves. These service courses may very well be the most important courses we teach—they may be the only association many students will have with biological sciences. It is important to realize that these people are going to be doctors, lawyers, merchants and congressmen. They are the people who are going to be asked to react to legislative proposals of great significance for education and science in this country. They should therefore get from these courses a background sufficient to react knowledgeably to future discussions of teaching and research in the biological sciences. If they are given good fundamentals in our courses, they will retain the knowledge.

A third question discussed here was when the introductory courses should be taught—whether in the first year and if so, the first semester or the second semester? Or should they be delayed to the second year and taught the third and fourth semesters? I would prefer to see the introductory course in biology delayed until the third semester, thus permitting the students to take courses in chemistry, math, and related subjects first, but they should not be delayed beyond the second year. To do this would make them competitive with the commodity-oriented courses the student will be required to take.

My preference is for a one-year course—two semesters that would probably provide introductory genetics, physiology, anatomy, and ecology as solid basic fundamentals to the commodity-oriented courses the student will take in the junior and senior years. In the latter case, I am thinking of courses in nutrition, plant breeding, animal breeding, physiology of reproduction, livestock management, plant management, and farm management services. We must keep in mind that the freshmen today, 18 years of age, will be 65 by the year 2015. So we must ask ourselves what kind of a world will he serve between now and then. We must try to educate him to understand and accept the changes that will occur. In fact, he must be equipped to visualize, plan, initiate, and direct the change. He is the architect in training for the future.

I feel that colleges of agriculture are the logical institutions to consider the ecological-environmental, natural resources-orientated biological sciences—the land, water, air, people, animals, plants and all of those things that go into making up our natural resources. They have been working with these in one way or another for a long time. They have certainly the knowledge and background that will enable them to continue effectively.

Food science and the application of biology to the satisfaction of the needs of man will, whatever the future, continue to be basic to his survival. Regardless of how far we may stray from our present concepts of food production, men will still have to eat. That means he must receive a combination of the basic food items—amino acids, minerals, vitamins, energy, bulk (all of which must be wrapped in an attractive package and tied with a pretty ribbon). Food does have an aesthetic value. Who will create the changes of the future? The biological scientist will certainly have as large a role as any other scientist. This knowledge will all be used to help provide a better place for man on this earth, to feed, to clothe, to house him so he can in turn enjoy more of the wonders of the future. Between now and the year 2000 we shall see dramatic changes—beyond that time we can only speculate, but let us remember that our students will either create the environment for change beyond 2000 or train the next generation that will create the change.

In agriculture at the present time approximately 7 percent of the graduating students return to agricultural production. To do so they must have a broad knowledge of the biosystems and interactions of plants, animals, soils, and climate. They must know why and how these systems work and be able to apply this knowledge to everyday problems. Successful agricultural production depends on four broad categories of knowledge: nutrition of plants and animals, genetics of plants and animals, disease and pest control, and management. Management is especially significant, because it really makes little difference what a person knows unless he can successfully apply it to everyday living and everyday problems. The farmer must make many decisions—he must select plants and animals for the stresses of high production and longevity, plants and animals that will yield a high-quality product economically. He must select feeds, seeds, and fertilizers that will perform adequately, at a price he can afford to pay. He must store and handle the produce in an economical, sanitary manner and manage his animals so they will yield the greatest return for the least cost. He must know how to control diseases and pests of both plants and animals.

The resources management specialist, like the agricultural producer, must know how to provide the proper social, economic, political structure so that maximum benefits can be derived from the inter-actions of man, climate, air, water, and scenery. This person needs a thorough grounding in ecology, sociology, humanities, and political science. The field includes outdoor recreation, resources development and alternate uses, park management, resources economics, soil and water science, conservation, and regional planning, to mention a few.

The agribusiness student will be the salesman for feeds, seeds, fertilizers, machinery, equipment, and other agricultural supplies. He will be the processor, the store manager, the transporter of foods, equipment, and services. He will be the agricultural reporter and edi-tor, the educator, the businessman—financing, banking, insurance, real estate, and consulting.

Finally, there is the graduate student—the one who eventually be-comes the teacher, the researcher, the director of the disciplines. Not all graduate students can be equally well prepared, which fact has led to a number of interesting schemes for remedying the situation:

- A common freshman year wherein all students receive a standard curriculum.
- A common freshman and sophomore year.
- A science-option to prepare the student for graduate work and a technology-option to prepare the student for business.
- Consolidation of courses to eliminate duplication and stressing of principles—e.g., consolidation of botany, zoology, and biochemis-try into departments or divisions of biological sciences; combination of dairy science, meat animal science, and poultry science into animal science; consolidation of dairy technology, meat technology, and food technology into food science.

G. FRED SOMERS

We are in a very fluid situation and the solutions to our problems are not as clear as we would like them to be. The course content and outlook of biology are rapidly changing, which leaves agriculture and natural resource technology in something of a quandary. They do not know where to turn, they do not see where the paths lead.

Let me remind those of you in agriculture and renewable natural resources technology that, like your colleagues in medicine and the health sciences, you have a primary concern for solving problems that are largely technological. I am sure you recognize that your technology is becoming evermore sophisticated, more demanding both in qualitative and quantitative terms. I think all of you recognize that the basic sciences of chemistry, physics, and biology, along with mathematics, are necessary to feed this technology.

We face a rapidly expanding continuum. Basic biology is rapidly becoming more molecular and quantitative at the very time when more attention is being given to the quantitative aspects of populations and their interaction with the environment. There is one dividend from this that we had not quite anticipated—i.e., the current emphases in biology are providing a deeper, broader, more substantial background for students in less time than formerly. We are now offering courses at the junior and senior level that were graduate-level courses only a few years ago. On the other end of this continuum, technology must convert these advances into practical solutions— solutions for problems that also seem to be expanding explosively.

In attacking curricular development there must be a thorough reorganization and a thorough change in point of view, a significant change in emphasis. As teachers in agriculture and natural resources approach the task, the content of their courses, the techniques they use, and the attitudes they assume resemble those of the organismic biologist of a decade or so ago. For example, plant production becomes more and more like plant physiology of former years. In this there is no conflict—indeed, it is an appropriate transition, for in some measure the biologist has already abdicated these areas.

There is need for a continuing, more intensive dialogue between agriculture and biology. In some measure when we ask which biology courses and how much biology agriculture majors should take we are begging the main issue. What we should do is to examine the total task before us, the whole continuum from the esoteric concerns of the so-called basic biologist to the most practical solutions that must be found by those who apply his knowledge. Doubtless we would find that at the ends of this continuum there is no conflict—our main task would be with the middle. We should be striving to make this interface as effective as possible.

Very likely the most important thing we can give our students is tools, hopefully sharp ones, to build the structure of their own careers. We might suggest how these tools can be used, but applications will probably be in ways that we do not yet see. We need to train our students to transfer knowledge from one discipline to another. It seems reasonable to expect that courses in the basic sciences have relevancy to problems in agriculture. This is a reasonable expectation but a more important goal would be to train our students to transfer basic knowledge to the problems they must face. This is a task that requires considerable insight and maturity.

To make suggestions of this kind is not sufficient. At least three concerns remain:

- Buying time in at least a few "pioneer" institutions for faculty to examine these problems in depth, with time and support necessary to work out solutions. Curricular matters, in the detail that must be our concern, cannot really be worked out at conferences, helpful though they may be. What faculty must do is sit down and consider these matters in depth. It seems to me this can be done initially at only a limited number of institutions. The patterns they then develop could serve as guides for others.

- "Retreading" faculty, periodically and in depth, to be supplemented by a continuing upgrading that might be provided by special seminars or training sessions. We might encourage professional societies to establish more and better training sessions in connection with their annual meetings—special symposia aimed other than at the specialist would be helpful. Symposia at professional meetings have been helpful in the past; perhaps we need more of them and need some reorientation of their objectives.

- Development of better texts and monographic materials, as necessary aids in supporting new approaches.

We need to look for new and imaginative ways to use the technology available. We need to look more carefully at such things as are available—unit concept films, audiotutorial labs, etc.—but we should prepare also for technology of the future. One that might be of considerable use would be shared-time instruction using computers. We should be prepared to throw aside our preconceived notions of course structure and content, to make as effective use as possible of available tools.

There has been concern whether or not we could accomplish all we would like within four undergraduate years. There must be a very real limit to the amount of material that can be handled by an undergraduate student, but students will rise to the occasion. We are not yet fully challenging our students, not yet fully using the potential of their high school training. We must start where the students are, recognizing that in some cases they are ahead of where we *suppose* they are. At various times in this conference it has been suggested that more use should be made of advanced placement and credit by examination—these seem desirable interim measures, but I would hope that we might examine more thorough approaches, using newer technologies, to advance the training of our students.

KEITH N. McFARLAND

St. Mark, 9:36, reads "For what shall it profit a man, if he shall gain the whole world, and lose his own soul?" Might we not paraphrase this to say "For what shall it avail the colleges, if they find the ideal curriculum plan, and do not have the student's commitment to learning?"

Everything that relates to curriculum organization, to content articulation, to methodology, must be measured by the single test, "What changes in the student were brought about?" An attitude of concern for the student and his growth is the frame of reference in which curriculum planning best takes place. And because the curriculum comes to life through the medium of skillful instructors, I have chosen to direct my remarks to the instructional process.

On the question of student input to educational planning, Task Force IX, a student study group contributing to a recent University of Minnesota Institute of Agriculture program review, expressed gratitude for "the opportunity to help mold change." The student report went on to say that "more student involvement can change the cry of 'student power' to 'scholar power.'" This student task force dealt with the outcomes of undergraduate professional programs and defined three objectives as being of basic importance—graduates should be technically competent; as a result of the activities and encounters provided by the program, they should be able to perform as productive citizens, apart from their professional role, and have interest in doing so; and they should have achieved an intellectual posture that supports continued learning. Of these, the third was deemed the most important, with this comment:

> Unfortunately, ... This quality has been the least heavily emphasized. At present it is not evaluated, and is not considered a prerequisite to a degree. In fact, research, as limited as it is in this area, suggests that few students acquire anything resembling an intellectual commitment. This is the largest, most encompassing problem in higher education. It must be dealt with before any significant progress can be made with other questions.*

To face honestly the question of student change brings a number of new elements into curriculum discussions. To plan content-oriented programs without reference to fundamental principles of learning is poor craftsmanship, indeed. Dr. A. M. Field, that grand old man of agricultural education, now long deceased, used to repeat that the role of any teacher is "to take the boy from where he is to where he ought to be." This may sound mundane, even trite, until one encounters a report like that in a recent issue of the *Chronicle of Higher Education*, where students in a well regarded midwestern liberal arts college described their curriculum as having "a relative lack of intellectual stimulation in the freshman year." The freshman curriculum was further described as being "customarily fragmented," characterized by a "general absence of perspective .... with respect to the liberal arts experience as a whole."† Such comment is grounds for searching introspection.

We face criticism of the way we help people learn. Administrators and faculty share responsibility for seeking solutions. It is my observation that in a productive educational setting, administrator or instructor behavior is largely determined by the individual's insight, imagination, and energy.

The following issues bear upon the quality of instruction, and hence, at least indirectly debate about curriculum.

● Rewards to staff come more quickly for performance in research or public service than in teaching. If this be so, consider the remarks

---

*Unpublished manuscript, Undergraduate Affairs, Task Force IX, Final Report to the Institute of Agriculture Long Range Planning Committee, 1968.
†Scully, M. G. 1969. Academic Innovation grows: Much of IR called "Fadism." *Chron. Higher Educ.* III(11):1,3.

made by Dr. O. Merideth Wilson (1966), President, University of Minnesota.*

Whether the role of the instructor is the transmission of codified knowledge or teaching the techniques for independent learning and research, one thing is absolutely essential. There must be an institutional concern for the faculty effect upon students. This is the first requirement. If teaching is honored by the institution, it will be cultivated; and its quality will in some way be assessed. And whatever hypothesis men have concerning the appropriate criteria for judging teachers, the question we should always be exploring is: How were the student learners affected?

● Students often feel that they have inadequate contact with senior members of the faculty. Advising contacts are at times hit-or-miss, large classes create feelings of anonymity, lack of proper facilities discourage informal talk.

● The examination system and schedule impose artificial restraints on students. Quizzes, mid-quarters, finals tend to categorize knowledge unrealistically. As one student put it "When one of the members of the faculty starts to work on an interesting idea, he stays at it. But students are supposed to be interested in chemistry at I hour, MWF, Math IV hour, TTHS, and soil science on selected afternoons."

● Teaching assistants may have but little feel for the classroom; some may not even speak intelligible English; resources to support instruction may be meager.

● The "service course" may be under stress because of the heterogeneity of student backgrounds, or because of ambivalence in objectives. The course may seek to serve as a first course for budding professionals and as an introductory experience, perhaps terminal, for others. The "common experience" courses or sequences may be received in a markedly different manner by students who are totally dissimilar in interests, aptitudes, and background experience.

● Experimentation directed to the instructional process is too infrequently encountered outside of the bureaus of institutional research or of colleges of education.

In an interesting article, Alvin Eurich suggested that four prime misconceptions stand in the way of change in instructional procedures.†

---

*Wilson, O. M., 1966. Keynote to annual meeting, American Council on Education, October 13.
†Eurich, A. C. 1964. The commitment to experiment and demonstrate in college teaching. *Educational Record* Winter:49–55.

1. Effective college teaching must be carried on in small classes.

2. There is a direct correlation between what the professor says in the classroom and what the students learn.

3. The student learns only, or best, when he is physically in the room with the teacher.

4. There is an inverse correlation between technology in teaching and individual freedom.

Soaring enrollments force innovation. The premises exposed to question by Eurich may on occasion have straitjacketed instructional approaches. Meager research efforts directed to the outcomes of varied approaches in instruction force us too often to positions and procedures based on opinion or grounded only in tradition. Too frequently objectives are defined in terms of content covered, rather than in performance terms, as in evaluation. How odd that in colleges of agriculture, where the research effort is so well organized and executed, such limited attention has been directed to research on effectiveness of instruction and program organization.

The point has been made that if the student is to command modern tools and resources, he must travel farther, and achieve more, than any group of students before him. He starts from complete ignorance. He moves with friends along the way, yet his progress is highly individual. He enters a post-high school program with widely varying competencies, revealing both trait and individual differences, depending upon his own nature and the circumstances of his earlier experiences. He is not a carbon copy of his contemporaries. He is reinforced by successes, highly vulnerable to failures. He has a limited preception of the many ways in which talent and energy can be applied, hence he is highly insecure. He is sensitive to what he sees as "major problems" of his universe, and resentful of apparent institutional disregard of them. He may have set unreasonable standards of performance for himself or for his institution. He needs to sense relationships, to have opportunities to practice or perform, if desired knowledge, skills and attitudes are to be established.

These factors bear upon course and curricular organization and upon instructor performance. Perhaps it would be helpful if curricular planners and instructors could step back on frequent occasion and ask themselves these questions:

• How *mature* are these students? Does their experimental background permit the use of the content or procedure I have in mind?

• Are they motivated? What is it about the way I am handling the situation that will make them wish to give it full energy and attention? Note the paragraph from the *Report of the Committee on Physics* that touches this point:*

While there is no doubt that physics is the best example of a science that has grown around the ideal of quantitation in application of theory and while it is central to our educational ideals in agricultural science to utilize a theory-based quantitative approach wherever this is useful, it is still more than likely that a traditional physics course taught without clear understanding of the agriculture student and his special needs may instead make him resistant to these very ideas and indeed antagonistic to related areas of mathematics and computer science that are becoming ever more vital to him.

- Is the timing right for this unit or activity? What are the special circumstances of this moment that give me teaching opportunities I should not miss?
- Do the students know why they are doing what they are doing, and how this activity relates to other aspects of their programs?
- Do they understand me? Am I using language, symbols, and illustrations that are pitched to their level of comprehension?
- Am I sufficiently sensitive to individual differences among students? Among freshman classes, reading ability ranges from perhaps the Grade 9 level to that of the top graduate category. How should this range of competence influence program organization and instruction?
- Is the instructional setting such that students are able to give full attention to the work at hand?
- Are my students passive or reacting? If learning is an action process, have I created a situation within which each student must become actively involved?
- And is the instructional program so arranged that the basic concepts are applied or practiced, in the interests of greater retention in learning?

Honest responses to these questions will help. The faculty man is a student. He keeps abreast of developments that bear upon his work and affect his performance. His knowledge takes on meaning because it becomes a part of his personality and influences his behavior. If we appraise the student's progress from a similar standpoint, perhaps the often encountered disparity between what appears to be the logical approach and that which appears to be psychologically sound, can be reduced.

*Preliminary Report of the Committee on Physics. Commission on Education in Agriculture and Natural Resources. Unpublished data.

APPENDIX A

Discussion Group Summaries

EXPLANATORY NOTE

At each of the four regional conferences, a substantial portion of the time was devoted to discussion groups. These were variously structured within a given conference, and from one conference to the next, but they fell into two categories:

- Groups having members of diverse interests, but addressing their attention to a set of questions formulated by the organizers of the conference.
- Rather homogeneous groups addressing themselves to a rather specialized aspect of the curriculum problem.

In summarizing these discussions, the comments of different groups in the first of these categories have been consolidated under the particular topic to which the comments are directed. No effort is made to identify either the members of the group or to indicate from which of the four conferences the suggestion emerged. Where commentary is limited, it can be assumed that that topic was not widely used in the series.

The reports of the special interest groups are also consolidated in those cases where similar discussions were held at more than one of the four conferences. In a number of cases only one such group was organized.

Where more than one discussion group formulated its response in very much the same way, both comments are retained, even though this results in a measure of duplication. There would seem to be added significance to a point arrived at independently several times by different groups.

TOPIC I

What are the broad areas of biological knowledge on which agriculture and natural resource technology is dependent, and to which all students in agriculture and natural resources should be exposed? Which of these areas need greatest emphasis?

Comment 1

The broad areas of biological knowledge that are needed by all college graduates include growth, reproduction, and their interaction with the environment as general properties of all living systems.

Comment 2

We define an agricultural or natural resource student as being anyone enrolled in a degree program in a college of agriculture, forestry, or natural resources. We feel that all students who are enrolled in such colleges should be taught some biological science, the minimum being a one-year course in biology.

Comment 3

Agricultural students actually represent six divergent areas, namely:

- Animal Sciences
- Plant and Soil Sciences
- Food Sciences
- Social Sciences
- Natural Resources
- Bioengineering

Recognizing this diverse interest we concur that one introductory two-semester course in general biology taken early in the curriculum would be preferable to separate courses in botany and zoology. Such a "functional biology" course should deal generally with all aspects of living matter. It should cover the six requirements of living organisms:

- Nutrition—metabolism and photosynthesis
- Gas Exchange—O_2 and CO_2
- Water and Homeostasis
- Waste Disposal—toxicity
- Control—organization, nervous system
- Reproduction—asexual and sexual

We consider this single course applicable to all university students regardless of major field and feel that it should demonstrate the natural interrelationship and interdependence of the animal and plant organisms.

Additionally, we feel that a sequence of: (a) cellular biochemistry, (b) genetics, (c) physiology, (d) comparative nutrition, and (e) ecology should logically follow for science, and possibly for technology, majors.

Comment 4

Broad areas of biological knowledge to which all students in agriculture and the natural resources sciences should be exposed include: ecology, cellular biology, genetics, physiology, reproduction, growth development, morphology, taxonomy, and behavior. A systems approach similar to classical ecology, but encompassing the response of organisms to the broad aspects of the environment, is important. Genetics and physiology are subjects affecting all of biology that may logically be used as a point of departure in describing the similarities and differences in living systems.

All students in agriculture and natural resources will benefit from having the same course, preferably one covering two semesters and including laboratory experience.

Comment 5

All baccalaureate degree students, whether they start at a university or a junior college, with majors in agriculture or natural resources should be exposed to a unified broad biology course (not a combination of botany and zoology), with particular emphasis on physiology, environmental biology (ecology) and genetics (classical, population and physiological). The course should include elements of developmental biology, morphology, reproduction, evolution and growth. The unifying theme of the course should be ecosystems, with emphasis on higher animals and plants.

For the two-year careers program the biology course should cover organisms and systems and should be of an applied nature. Admittedly, finding biology instructors for the applied type of course will be difficult for the two-year schools.

Beyond the beginning course in biology, the diversity of individual student needs, interests and curricula requirements become so diverse that delineation of certain courses as essential for all becomes impracticable.

Comment 6

The broad area of biological knowledge upon which natural resources curricula is based is most certainly, and primarily, ecology. Other areas of biology—e.g., molecular and cellular biology, genetics—can be adequately covered in an introductory year. To most natural resources areas, physiology is a necessary biological component. Although there is unity at the cellular level, students in the natural resources areas deal with organisms, and there is a difference in organ physiology between plants and animals. Thus, separate plant and animal physiology courses will be part of forestry and wildlife curricula, respectively. Physiology is not essential to resource students whose interests relate solely to water or soil.

Although not strictly biological, soils—their composition, their microbiology and the biogeochemical cycles that involve this phase of the ecosystem—are also an area of study to which students in all natural resources curricula should be exposed.

These courses, along with the ancillary courses in math and the physical sciences, form a common core in the first two years of natural resource curricula—no matter what the resource is and no matter whether the emphasis be strictly scientific, management, or business oriented.

Comment 7

The broad general area of biological knowledge required of all students in agriculture is knowledge of the cell—its structural components, functions, nutrition, and replication. Included in this would be the identification and function of cell membrane, mitochondria, ribosomes, and the nucleic acids and their role in protein synthesis— how groups of cells are structured into organs, and how plants or animals interact in various populations. The greatest areas of importance would stress the function of the cell in relation to important biological processes—e.g., photosynthesis, fermentation, pathology. In addition, emphasis could be place on the interaction of populations of cells, organs, and plants and animals—in other words, the specific relevancies of biological processes in agriculture. This information would be included in a two-semester course sequence offered during the freshman year. It assumes a high school background in chemistry and mathematics of one year each. Deficiencies in these areas would seem to dictate that this course be deferred until the sophomore year, allowing the chemistry and mathematics requirement to be met during the freshman year in college.

Comment 8

There is a definite identifiable body of knowledge that can be prescribed as the minimum basis for an adequate understanding of modern biology, regardless of the future specialization of the student, be it agriculture, biology, medicine, education, or some other field. The basic body of knowledge should be the same for all the above groups, and include:

- Cell Biology
  Structure and function
  Plant and animal cells
  Procaryotic and eucaryotic cells
  Osmosis and diffusion

- Molecular Biology
  Elementary biochemistry
  Intermediary metabolism
  Photosynthesis
  Protein synthesis
  Structure and function of nucleic acids

- Genetics
  Mendelian genetics
  Biochemical basis of heredity

Population genetics
Genetic change
Evolution and speciation (introduction)

- Ecology
Ecosystem
Ecological factors—biological and nonbiological
Populations and population dynamics
Practical ecology—birth control, etc., environmental
    contamination, conservation and natural resources
    management
Manned space flight and space biology
Oceanography

- Structural and Functional Organization of Living Systems
Gaseous exchange
Nutrition
Fluid balance and excretion
Reproduction, life cycles and development

- Evolution and Diversity of Life
Patterns of evolution—animals, plants
Mechanisms of evolution
Classification—principles

Comment 9

All students in biology, including agriculture and natural resources, should be exposed to the same broad areas of biological knowledge. The core of courses offered in each institution for students in the biological sciences should incorporate basic concepts common to all living things. This core should include discussions of (1) the diversity of living organisms, (2) their evolutionary development, (3) their phylogenetic relationship, (4) their structure and function, and (5) their response to their environment.

We subscribe to the philosophy that the principles encompassed by the generally accepted concepts of cellular, organismic, and environmental biology should constitute minimal offerings in a biological sciences core.

The sequence of presentation of these concepts, emphasis and the development of finite courses should be left to the discretion of individual departments and installations in relation to the needs of the students and the administrative organization of the institution.

Comment 10

Topics from every major category should be presented to all students in agriculture and natural resources early in their academic experience. The relative exposure to each category should be as follows:

Rank	Category	Relative Emphasis
1	Ecology	28%
2	Genetics	14%
3	Evolution	10%
4	Physiology	10%
5	Development and growth	9%
6	Reproduction	8%
7	Morphology	8%
8	Cell biology	7%
9	Taxonomy	6%

These topics should be presented in an integrated approach to biological systems, using a balanced selection of organisms as examples. The order and manner of presentation of these topics will reflect the nature of the individual institution; it should include those recent developments not subject to rapid obsolescence.

TOPIC II

In what ways are the biology needs different for the various majors in the college of agriculture? In various options: business, technology, and science?

Comment 1

Students who are in nonscientific specialties such as agricultural business and economics, agricultural engineering and home economics, should take a special course in biology that is designed to suit their needs as a terminal course. This course would encompass cellular, physiological, ecological, and genetic relationships in biology. Molecular biology would not be stressed here.

Students in the science-oriented programs of agriculture and natural resources should take a one-year biology lab course. In addition, they should take a course in genetics, as well as one course each in physiology and ecology.

Comment 2

There is an obvious difference in the biology needed for the various options offered in the modern colleges of agriculture. However, to devise a truly effective program for the individual student, it is of paramount importance that an advisor qualified in a given area work closely with the student to devise a program based on the student's own aims and purposes. In general, those students who are biologically oriented should be encouraged to take as much biological science as is available to them on the campus. Those who are oriented toward economics and other social sciences, as well as agricultural engineering, should be similarly encouraged to take additional courses in biology over and above the initial one, unless there is indication that other courses in general education would be more advantageous.

Comment 3

Following the introductory course, students in agricultural economics and socially oriented subjects, as well as agricultural engineering should take about six additional semester hours in biology-oriented courses. These latter courses may be either in basic biology or in such applied biology as plant or animal production.

Students majoring in applied areas of biology in agriculture and natural resources as a background for such applied courses as animal and plant breeding and management that are to be taken later in their major departments. Biochemistry appears to be adequately handled either in specific courses or as part of applied courses.

There seems to be real difficulty in bridging the gap between such fundamental courses as molecular genetics and such applied courses as plant breeding. The situation seems to offer an important opportunity and challenge to those teaching applied courses to use modern teaching techniques—e.g., programmed instruction on closed circuit television—to help the students make the most of their time and ability in dealing with material that has usually been assumed to belong to the basic course.

Students in technology and science programs should take similar—

or even the same—courses in basic biology, the science majors then taking additional courses in math, chemistry, and physics. Technology students will ordinarily take more applied biology and management courses than do science majors.

Comment 4

There is need for frank, concerned dialogue between those in the agricultural sciences and those in biological sciences. Even so, it is improbable that a single course could satisfy all requirements for all students.

A basic animal or plant science course in the first year is important to maintain interest and motivate the students, as well as to demonstrate direct applications of the biology course to the major fields of endeavor.

In most universities, adequate courses applicable to the needs of biologically oriented agricultural and natural resources students are available. When needed courses are not available, they should be established in whatever department has the greatest competency to teach them.

Comment 5

The introductory course would be appropriate for all majors in the various options in agriculture. Other courses that should be offered by departments of biology for students in technology and science include: genetics, physiology, microbiology, parasitology, limnology, ethology, and ecology.

Comment 6

We recognize a need for liberalization in the undergraduate curriculum—not only increased exposure to specific courses in mathematics and the sciences, but more opportunity for courses in the social sciences and humanities. There is, however, real doubt whether courses as currently offered in the social sciences and humanities do indeed increase the facility of a resource manager in solving "people problems."

Liberalization of the undergraduate curriculum can mean that preparation for professions in the natural resources will require a fifth year, or master's degree program, in which the student will be in-

volved in advanced, rigorous ecology as well as practical technique
courses.

TOPIC III

What can be done to develop in students of agriculture and natural
resources a greater consciousness of the relevance of biology to agriculture and natural resources?

Comment 1

The relevance of molecular and cellular biology to students in the
natural resources might be enhanced if the biology core began with
an ecological orientation. Yet, it is still the responsibility of the instructor of any given course to illustrate its relevancy. Thus, he must
know to whom he is relating. Specific relevancy not only of biology
but also the physical sciences and mathematics to resource curricula
could be made the responsibility of the major department and could
be implemented by having a continuing course that emphasized pertinent problems and methodology.

Comment 2

The most effective means for impressing students in agriculture with
the relevance of biology to agriculture is to stress ways in which biological processes relate to conditions in agriculture with which the
student is familiar.

TOPIC IV

In what ways are the biology content needs different for the various
majors in agriculture and natural resources, such as horticulture, food
science, forestry, and agricultural economics? In various options, such
as business, technology, and science?

Comment 1

The group believes that the minimal needs of students in all phases of
agriculture are the same, although this would not necessarily apply to
two-year courses in technology.

Comment 2

Beyond the introductory level, the needs of a student in agriculture and natural resources begins to diverge from those of the biology major. The level at which divergence occurs, and its extent, depend upon the program and goals of the individual student.

Opportunities should be provided for "in depth" study in each category. The number of categories and the depth of study depend upon the student's interest and goals, e.g., the pregraduate degree students in biologically oriented areas of agriculture and natural resources science would take the same courses as the biology major, whereas the social science-oriented students might take no advanced courses.

TOPIC V

How can the knowledge that the student gains from the various courses in the curricula be integrated so that he is prepared in the best way possible to perform professionally?

Comment 1

Integration occurs when material from core biology is used in courses for agriculture majors. This implies that teachers in agriculture must become better acquainted with what is being taught in the basic biology courses if they are to use this information effectively. In addition, it is imperative that those teaching the core biology science courses be informed as to the uses being made of their material. Thus communication is encouraged.

Comment 2

Integration of knowledge from various courses must be accomplished by each student on his own. Faculties should assist the student in this endeavor at every point. Orientation courses, advising, work experience, use of appropriate examples in applied courses, senior synthesis courses and tutorial courses are all useful techniques. Students should be continuously motivated to understand the need for, and the techniques of, successful integration of knowledge.

Faculty must be well informed on material being used in courses other than those they are teaching if they are to be successful in assisting students in the integration of knowledge. There are many formal and informal ways by which faculty members can become so informed. Most faculty members and administrators are aware of these techniques; it is important that they have the motivation and the time to take advantage of them.

Comment 3

The discussion centered on student advising—the needs and the means. While there was no consensus as to the number of students that an advisor might capably handle, the group was convinced that student-faculty contact through a "good" advising system is urgent and that advising can play an effective part in the integration of both courses and knowledge, particularly at the junior and senior levels.

Other suggested means of integration are

- A core curriculum so designed that courses build on preceeding ones, as in a professional program.
- Use of electives in the senior year, so that a student may select courses that allow for experimentation, for actual application, and for search and examination—as in seminars.

Comment 4

The group felt that integration might be accomplished in the following ways:

- Honors programs, special topics or courses of the senior-theses type, all of which involve close faculty-student dialogue—but can be very demanding of faculty time.
- Seminars, field trips and guest speakers—a more flexible approach that has, of course, the inherent problem of coordination characteristic of team teaching.
- Senior courses, such as ecology, management, policy, economics—keeping these courses as modern, relevant and integrated as possible.

In addition, it was felt that the faculty should be integrated in its philosophies of teaching; continual educational opportunities for faculty members was stressed.

Providing actual work experience is a highly successful way of helping the student to relate to his formal studies. This experience may be

afforded through placement training or by laboratory and field contacts.

The student should be encouraged to think broadly, to meditate
and to philosophize on the implications of his knowledge.

Comment 5

The program must be so constructed objectively so the student can
make use of the knowledge gained.

• Seminars should utilize advance students on an all-campus basis,
developing maximum cooperation among disciplines with a view to
producing experts in areas that cut across related subject matter.
Overscheduling should be avoided, allowing adequate time for free
and open discussion. Finally, seminars should make greater use of
senior research projects, of special-problems classes for outstanding
advanced students, and of formally organized interdepartmental
courses.

• Efforts to make the student aware that his specific program of
courses has a general bearing on all biological problems facing man
should not be delayed until he is ready to graduate. Indeed it may be
desirable to require all freshmen to take a year seminar that deals
with biological, sociological, political issues, emphasizing the interrelationships and that individual segments of knowledge are useful
only if they are integrated and brought to bear on solving the problems facing modern man.

• It would be well to utilize extensive integrating field study tours
in an unfamiliar environment as a way to encourage analysis of a
complex agriculture or natural resource situation. Many students
accumulate bits and pieces of information from each course in their
curriculum but are unable to relate it to an "on the spot" situation.
We advocate field study tour during the summer months, or between
semesters, that requires students to integrate the factors that influence a complex agricultural or natural resources situation. Special
guidance and orientation, where needed, should be provided by
specialists. There needs to be an integration of industry, business,
social aspects, population, fundamental and technical knowledge, and
the natural physical environment if students are fully to understand
a given situation. Courses attempting such an integration of material
may be offered by an individual department, by departments cooper-
atively, or on a multi-university basis.

● Whenever possible, prerequisites for undergraduate courses should be eliminated, and every effort made to integrate prerequisite information into subsequent courses. At times, this may necessitate professional upgrading of some faculty members.

● An important integrating step must be to break down the wall that separates agriculture and liberal arts. After all, most colleges of agriculture are designed to offer a liberal arts education and are in a position to fill an important service role to the entire university.

TOPIC VI

Should biology courses in the two-year technical program be designed so that they are transferable to baccalaureate programs, or are the requirements of a technical curriculum such that the biology courses should be different from those in the bachelor's program? If they should be different—how?

Comment 1

Biology courses that are transferable to four-year programs cannot be designed for terminal two-year agricultural students. However, terminal biology courses should be taught in two-year units with the same approach as for four-year programs, to the extent that they can be worked into the program at the local level without endangering the student's eventual employability. Biological science courses offered in nonterminal two-year colleges can be designed so that credit is transferable to a four-year program. To facilitate this transfer, they should be designed as if for the four-year students.

Comment 2

If a two-year program is terminal, its biological courses need not be designed to be transferable. If, however, it is a preprofessional program, its courses should be so designed. In the final analysis, transferability is determined either directly or by examination in the department granting the credit.

Comment 3

Biology courses in two-year technical programs should not be transferable to baccalaureate programs, but biology courses offered in junior colleges as part of a regular transfer program would be acceptable.

Comment 4

Two-year career programs are occupationally oriented and are responsible for turning out a student who is technically trained, confident of his training, and employable at the time he receives his certificate. Obviously, a sizeable percentage of these students may transfer to a four-year program upon completion of the two-year program. But the design of the curricula, and the content of the courses offered, are essentially to be determined by the needs of the student going directly into employment.

Comment 5

There is a real difference between the two-year technical and vocational curricula and the four-year curriculum. They have mutually exclusive tasks. The two-year technical course is directly related to a job; it meets an immediate need. The four-year curriculum must be directed toward professional roles that may not yet even exist, hence it must be concept—rather than application—oriented. We feel also that a course in ecology should be part of *all* majors in agriculture.

Comment 6

Biology courses in the two-year technical school should not be specifically designed for transfer to B.S. degree programs, but rather primarily for direct application. There is no objection to these courses being accepted for transfer, but it would be an incidental feature. The fact that many two-year students successfully transfer to B.S. programs testifies to the caliber of the course material.

Comment 7

Biology courses in two-year colleges must fit the objectives of the school. If it is to train technicians, only a minimum of basic biology courses need be offered, although, because of rapid changes in technology, some biological principles will be helpful in enabling graduates to learn new procedures. If the objectives of the school are to educate the students for transfer to four-year schools—and this appears to be a growing objective of some technical programs—basic biology courses similar to those offered in four-year schools should be offered.

TOPIC VII

Are sourcebooks needed to improve instruction in the biological sciences for students in agriculture and natural resources?

Comment 1

Sourcebooks would be very desirable if broadly defined as a general fund of information specifically relating biology to agriculture. This source of information should be available to instructors of biology on a purely voluntary basis. As an alternative to making sourcebooks, members of agricultural staffs might participate as assistants in biology laboratories with responsibility for demonstrating biological principles in specific agricultural situations.

Comment 2

No need for a sourcebook to identify the relevance of biology to the natural resources.

TOPIC VIII

How can the recommendations of the discussion groups be implemented?

Comment 1

The general recommendations of the discussion group might best be implemented by:

- Joint appointments between departments in core and applied fields.
- Development of intra- and intermural dialogue, including the exchange of course syllabi or outlines, so that faculty might orient one another as to course content and offerings.
- Expansion of agricultural courses in breadth to win university-wide recognition, appeal and acceptance—e.g., courses in man and his environment; world food supply; genetics and world population.

● Maintenance and improvement of student counseling and advising systems, so that more individual attention could be given to the specific needs of the student—a very critical aspect of the total education of the individual.

● Increased recognition of superior teaching and advising as key objectives of the total university program.

Comment 2

Group members can be most effective in implementing recommendations by transmitting their ideas with their colleagues on the home campus and by closely scrutinizing their own activities to determine their relevancy and effectiveness.

Comment 3

It is desirable to implement recommendations while the ideas and the stimulation are still fresh. Efforts by attendees to promote and activate the ideas generated are the best means of implementation.

Comment 4

The following are suggested ways of implementation:

● Discussion with the staffs of participants' home institutions.

● Discussion of specific needs with members of the departments of biological sciences.

● Development and assembly of reference materials.

● Experimentation with new approaches to teaching—alteration in course sequence, team teaching, synthesis of "interest" courses.

● Providing, in core biology courses for assigning students in the various disciplines to divisions of the core that emphasize areas of specific and different interests, and facilitating movement in and out of the core with a minimum of prerequisite restriction.

● Instructors in biology and agriculture should receive encouragement from their administrators to improvise and cooperate in teaching courses in their respective areas.

Comment 5

Recommendations can best be implemented by active participation of faculty in individual institutions under the continual stimulation of national groups such as those sponsoring the conference.

Administrators at local institutions can be helpful by creating the proper atmosphere and supporting faculty efforts.

Conferences in which all disciplines and departments involved participate in the decision-making process can be important.

Successful implementation will require recognition of mutual objectives and a feeling of understanding and confidence among the different faculty and disciplines involved.

It is important that those who propose changes make certain that they are soundly based, well thought out, and will result in real benefits to students and to society.

Recommendations can be put into effect more quickly by techniques that bring about increased contact among faculty and students in different departments and disciplines. Exchange of lectures in courses, luncheon meetings, joint seminars, and formal faculty meetings are means to this end.

There is a strong challenge to faculty in applied areas to become familiar with both the basic and applied areas of science, and to take the initiative in developing programs in cooperation with other disciplines that will result in the best education possible for students coming under our responsibility.

Comment 6

Implementation of the above suggestions could be encouraged by the following:

- Strong adherence to prerequisites for advanced courses should be followed, and prerequisites established if they do not now exist.
- All-day meetings at local institutions should be funded, to enable faculty members from the biological and agricultural sciences to discuss implementation of the basic biology core courses at local levels, recognizing that the opportunity already exists at many institutions.
- Whenever employment within the respective agricultural and resource disciplines during the undergraduate experience is feasible, it should be encouraged.
- Administrators and others should give more attention to establishing a system whereby faculty in agriculture and natural resources may participate in continuing education.

APPENDIX B

Two-Year and Four-Year Curricula

GENERAL COMMENT

Open door policy of many junior colleges does not mean there is an "open course" policy—students must make acceptable scores on placement tests or show satisfactory achievement in preparatory courses to get into many transfer courses.

A look must be taken as to what graduates need by way of preparation; not all agriculture majors need the same biological science preparation. Therefore, two or more levels of biology should be available. We too often try to make the student in our own image.

If one accepts the premise that colleges are in business to help students, they must do what they can toward this end. Undergraduate programs must take into account the fact that many students change majors one or more times while in college—in some cases over 70 percent of the sophomore students change majors at least once. Those who do change majors will encounter some difficulties.

Some students will use general education biology as an introductory course in biology sequence, for others the course will serve as an elective.

Articulation between junior colleges, state colleges and universities is important and must be improved if students are to flow smoothly through the institutions.

● Preparation, drive, the teachers he has had—all these influence the transfer student.

● Articulation must be worked out, and with more than just the information found in the catalog.

In view of the degree of interaction that takes place between junior colleges and universities, general problems in one institution are problems for all.

More and more upper division students are now coming, and will continue to come, from two-year institutions.

TOPIC I

Should the same kind and amount of biology be taught to (1) transfer students enrolled at junior colleges and (2) lower-division students of the 4-year colleges and universities? If different, how?

Comment 1

Transfer students enrolled at junior colleges and lower-division students of the four-year colleges and universities should receive the same kind and amount of study in the biological sciences.

Comment 2

The same kind and amount of biology should be taught to transfer students enrolled at junior colleges and lower-division students of the four-year institutions.

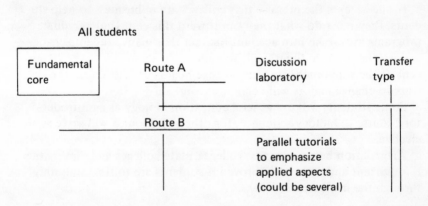

Two-year colleges should have special programs for those students who terminate their formal education at junior college level. Transfer students should have either a general biology for nonbiology majors, or a more detailed rigorous course for majors in biology and in certain agriculture and natural resources areas. If only one transfer biology course is offered at the junior college, it should be the more rigorous course that biology majors would be expected to take. There is no support for a distinction in kind, or amount, of biology for junior college (potential) transfer students and the lower-division students in four-year institutions. There is, on the other hand, good reason for a separate biology course for general education students, thus providing at least a two-track system.

Course for majors would be given in depth. Suggestions for general education program include emphasis on man and his environment and an interdisciplinary approach.

Comment 3

There was general agreement that the courses should contain the same subject matter, recognizing that it can be taught in many ways. There was also a feeling that schools generally should come to an agreement as to the unit value given to courses, although it is acknowledged as a secondary issue. Four-year institutions should unify their courses, individually and collectively; the two-year institutions would in turn then have a guide to follow in organizing their own courses.

TOPIC II

Should biology courses in the two-year technician program be designed so that they are transferable to baccalaureate programs, or, on the other hand, are the requirements of a technical curriculum such that the biology courses should be different from those in the course parallel program? If they should be different—how?

Comment 1

Biology courses in the two-year technician program should be designed so that they are transferable to baccalaureate programs, if such courses do not detract from the vocational-technical objectives of the individual junior college.

Comment 2

Separate biology tracks should be available in the junior college for terminal students (ex-technical programs) and transfer students. Because the terminal student enters an occupation promptly and has different interests and ability, the biology course for him should vary from that of the transfer student in depth and emphasis on applied aspects.

Comment 3

Not only are there technically oriented students and the professional students, but also nonmajor students interested in a general course that will show how biology fits into various aspects of life.

The course designed for the technical student should omit portions dealing with biochemistry and comparable topics—it should stick to basic biology. This does not imply a "watered-down" course, but is rightfully termed an abridged course, quantitatively shortened, but not qualitatively lessened.

There should be different courses, each of which may be transferable if the appropriate purpose is kept in mind. There are, then, three objectives: (1) a two-year technical training program, (2) a program for the professional major needing a comprehensive, intensive course, and (3) a program for the nonmajor interested in a general course.

TOPIC III

Assuming that, ideally, different types of biology should be available for technical and transfer students, how can the biology instruction be designed to best meet the needs of all students if it is not possible to separate technical and transfer students?

Comment 1

When parallel programs for technical and transfer students cannot be offered, it may be necessary for the college or university to combine sections. In such cases, the professor should examine student backgrounds and departmental majors or minors so that he may incorporate experiences that are appropriate to the total class population.

Comment 2

If only one biology is offered—this was not considered a frequent operational necessity—it was agreed that the single course should be of the transfer type. Students might take general education or the major course and, in addition, a one-unit parallel tutorial.

Comment 3

If it is not possible to have separate courses, then there must be an integrated course that covers all basic principles of biology. This will require an unusually well-qualified instructor or a well-organized and supervised team of instructors. Whether the course is taught by one instructor or a team, there must be proper coordination and communication between department members.

TOPIC IV

How can problems in transfer of biology credits between institutions, especially if varying amounts of biology are required, be reduced or avoided?

Comment 1

Some junior colleges may be unable to comply with all requests for biology due to a lack of financial support or facility. This situation will make it necessary for each senior institution to have an evaluation committee with responsibility for advising transfer students as to biology in which they are deficient. Continuing efforts should be made by senior college personnel to keep administrators, advisors, and students of junior colleges apprised of the senior college course requirements.

Comment 2

The problem can be avoided by unifying course requirements in the four-year institutions, so that the two-year institutions can more easily design their own courses.

The acceptance of a transfer course is based primarily on their content and organization. The course should not be labeled as "terminal" or "nonmajor"; numbering systems mean nothing.

There must be *complete* coordination between and within both the two-year and four-year institutions.

One must keep definitely in mind the various goals that students have. Ideally, different courses should be offered, but, in any event, courses must not become "watered down." Subject matter content is all important. Course titles and numbers are often unrealistic.

Comment 3

An attempt should be made to reduce or avoid transfer problems between institutions through:

● General conferences; articulation with junior college; regional exchange.
● Sharing of curriculum information.
● More advising and less counseling at the junior college level—the percentage of transfer students will increase greatly, and they vary greatly as to (1) ability of student, (2) selectivity, and (3) prior counseling and advising. Agreement will have to be reached on freshman-sophomore curriculum.

The title of the course is important. If names and numbers are brought into conformity as much as independence of operation permits, it would be helpful. Instruction impact is most important aspect of instruction, improved by help in curriculum content and progression.

Articulation, such as that carried out by Arizona Biological Conference, brings together junior colleges and four-year institutions. Other states can benefit from experience of California in articulation and thus avoid pitfalls. It is wise to articulate early, regularly and at all levels.

APPENDIX C

Plant Sciences

GENERAL OBJECTIVES

• To provide a unified body of basic biological knowledge as a prerequisite for discipline-oriented training in the agricultural sciences.

• To provide a flexible biology "core program" that will suit both "technological" (agribusiness) and "science" curricula.

No distinction should be made among plant, animal, and natural resources students at the early stage of their training, nor between technologically and scientifically oriented students. The provision for flexibility allows adequate freedom for all options.

The use of a "core" program allows duplication to be avoided in agricultural courses. Perhaps more importantly, it provides a common base of biological knowledge on which to build specialities.

RECOMMENDATIONS

We recommend that the core biology program include in-depth information in the following areas under the general term "principles."

- Chemical basis (molecular aspects of biology).
- Structural basis (cellular and multicellular aspects of organization).
- Energy basis (synthesis, degradation, and transport mechanisms).
- Control basis (homeostatic mechanisms at the cell and organism level).
- Differentiation, growth and reproduction (patterns in unicellular and multicellular organisms).
- Genetic basis (transmission of genetic information at the cell, organism, and population level).
- Systematic and evolutionary basis (classification or organisms with the through time).
- Ecological basis (concept of ecosystems at several levels of complexity).

The principles must be illustrated in laboratory exercises appropriate to the subject matter. In these, instrumentation and organisms should be handled by the students. Certain aspects may be fitted to audio-tutorial systems; others to more traditional methods.

Diversity in the biological program will show up in two ways: (1) in reoriented courses in the student's specialized field as he moves from the core program, or (2) by greater in-depth treatment of the biological principles as they relate to the student's special field.

IMPLEMENTATION

Diverse routes to provide the core of biological information are desirable at this stage; no one route is necessarily best for all institutions.

The core biology program should be incorporated into the curricular offerings as soon as possible. Concurrently, course reorientation must occur in the various disciplines.

The minimum core program must be implemented at the junior college implementation, audio-visuals, audio-tutorial, TV tapes, slide sets, and other teaching materials should be developed. As these become available at various institutions, they should be cataloged by some central, nonprofit agency that would take the responsibility for providing "availability information" to all institutions.

APPENDIX D

Quantity of Biology

TOPIC I

To what extent—e.g., number of quarters or semesters—is a core program in biology feasible for all students in agriculture and natural resources?

Comment 1

There was general agreement that a core program in biological science is feasible for all students in agriculture and natural resources. It is proposed that this core consists of a one-year's sequence (3 quarters or 2 semesters) of 15–16 quarter units, or 10 semester units, and include a laboratory. It should include cellular, molecular and organismal biology, which would start the 3rd quarter or 2nd semester of the freshman year.

Comment 2

The consensus was that two semesters or three quarters of beginning biology courses in a core program of biology are feasible and desirable for all students in agriculture and natural resources. Cellular, molecular, organismal and population biology should be covered. Advance placement should be encouraged. One of the advantages of a uniform first year of biology is that students may more easily shift majors as their interests and goals change. There was no agreement as to whether biology should be in the freshman or sophomore year. Some departments of biology choose to offer the beginning biology courses in the sophomore year, following a year of chemistry and mathematics as freshmen. Since the introductory courses in biology are considered valuable prerequisites for courses in agriculture and natural resources, some of the participants felt biology courses should be offered in the first year. The basic unresolved question is whether the introductory biology courses should be taught at a level requiring college chemistry, not whether training in freshman level chemistry, *per se*, is needed by students in both biology and in agriculture and natural resources.

Comment 3

There was a wide diversity in the group as to opinions and experience with curricula. The size of institution appears to have a significant influence on course offerings. The following was generally accepted as a definition of core curriculum:

A one academic year introductory course open to all students, followed by a minimum number of advanced courses acceptable to the various disciplines in agriculture and natural sciences—usually extending over a one or two year period.

The consensus was that the first year core program could be so designed that it would be acceptable to the various disciplines, regardless of the student's career objective.

The importance of complete course descriptions in the college catalogs was stressed, in order to avoid repetition in course sequence.

Advance placement was recommended as a means of avoiding repetitious material for the better students.

TOPIC II

Should students seeking only a B.S. degree have the same amount of biology as those who are planning to do master's and Ph.D. work?

Should the amount vary by *major field* and by option (business, technology, science)? If there should be differences—what are they?

Comment 1

The group was in agreement that there should be no difference in the basic core courses regardless of whether a student is seeking only a B.S. degree or is planning to do graduate work. Beyond the core program the various curricula should be flexible enough to allow, and encourage, the prospective M.S. or Ph.D. student to obtain training in more depth, particulary in the quantitative aspects of biology and also in mathematics and chemistry.

Comment 2

No, following an integrated first year of biology courses, students should be steered into varying amounts and kinds of biology, depending upon their goals. This counseling will vary among options. Obviously, students in agricultural economics do not need the same amount of biology as do students in the plant and animal sciences. Marked differences exist between needs of students within the plant and animal sciences—for example, prospective graduate students in soil sciences need relatively more training in the physical sciences and less training in the biological sciences than do prospective graduate students in animal breeding.

Comment 3

In the opinion of the group, first year students usually do not have firm enough career plans to justify varying the "core program" to any appreciable extent. Following their first year, their choice of biological courses, with the assistance of their advisor, should be so structured as to complement their program in the area of major interest.

TOPIC III

How much emphasis should there be on biology at the different levels—lower division and upper division—in terms of a proportion of the curriculum at each of the two levels? (Assumption should be made as to major field and option being discussed.)

Comment 1

Because preparation in depth in biological science is a requisite for success in agricultural curricula, major emphasis should be at the lower division level.

Particular emphasis should be given ecology and genetics, though not necessarily only in the lower division.

General ecology offerings for agriculture majors should deal with ecosystems and their relevance to man.

It was generally agreed that courses in general physiology and organismic biology would be desirable for upper division students in plant and animal sciences.

Comment 2

The proportion of biological science requirements allocated to lower and upper division must be considered in the light of differences among majors and among options within majors. All options have approximately equal needs in biology in the lower divisions; in the upper divisions, they do not. Plant and animal science, *per se*; plant pathology; and range, wildlife, fishery, and forest resource management curricula have heavy upper division biological components. Agricultural economics, business, forest business, agriculture, and forest engineering and wood utilization curricula have relatively less upper division biology courses.

The following comments were made by individual members.

* Curricula should be seriously re-evaluated, appraising the need for, and content of, all courses.
* Encourage formation of a committee on each campus, composed of faculty members from biology, agriculture and natural resources, to study content of beginning courses in biology.
* Graduate students should be enlisted to provide advice concerning content of undergraduate courses, since their exposure is more recent than that of most staff members.
* Faculty in agriculture and natural resources should become better acquainted with teachers of biology courses and with the reasons why biology professors believe courses should be as they are.

- Increase and encourage competence in teaching, with in-service training; give graduate students some experience in teaching.
- Cease making an issue of "relevance" in introductory biology, and credit students with the substantial perceptiveness they possess.

Comment 3

The concept of nonterminal, unified courses in biology at equivalent for all up to the point where they spin off—wherever this occurs— was accepted. There was no agreement with regard to specific majors and options as to when this might occur, nor as to the proportion of the curriculum to be devoted to biological science at the upper- division level. The following table (p. 150) suggests a number of pos- sible alternatives.

Suggested Core Program in Biology for Students in Agriculture

---

Year

Freshman   Fundamentals of Biology
                            (Taken 3rd Quarter of School year-after chemistry)
                            Oriented on cellular and molecular level
                            6 quarter units (4 lectures, 6 hrs lab/wk)
                            Prerequisite - 2 quarters of College Chemistry

Sophomore   Biology of Organisms
                            Structure - function, growth, development,
                            physiology, reproduction, phylogeny.
                            Integrated 2 quarter sequence
                                5 quarter units (3 lectures, 6 hrs lab/wk)
                                5 quarter units (3 lectures, 6 hrs lab/wk)
                            Prerequisite - Fundamentals of Biology
                            Corequisite - College physics

                                → *Curricula requiring minimum 2 yr Core*
                                Agricultural Econ. and Bus. Mgmt.,
                                Agricultural Engineering, Home Economics,
                                Textile Science, Dietetics and Nutrition, De-
                                sign, Child Development, Family and Con-
                                sumer Science.

Junior   Biology of Populations
                            Environmental Biology
                                5 quarter units
                            Prerequisite - Biochemistry

                                → *Curricula requiring preceding sequence*
                                Preforestry, Range Mgmt., Food Science, Soil
                                Science, Water Science, Atmospheric Science.

Senior   Genetic Biology
                            5 quarter units (Prerequisite - Biology of Popula-
                            tions)
                            Selected courses in areas basic to major field

                                → *Curricula requiring preceding sequence*
                                Agr. Education, Animal Physiology, Animal
                                Science, Nutrition, Entomology, Preventive
                                Medicine

APPENDIX E

Chemistry, Physics, and Mathematics

GENERAL CONSIDERATIONS

Comment 1

In view of the need for more chemistry, physics, and mathematics in the undergraduate curriculum, it seems necessary to find ways to introduce such courses into an already crowded curriculum. To maintain a four-year bachelor of science program, already-existing courses must be consolidated, and extraneous or superfluous material eliminated.

Certain prerequisite courses could be abolished. As high school preparation becomes better, many introductory courses could be eliminated, and those used for agriculture students during the freshman year must contain significant depth if they are to be justified on subject matter alone.

The group recognized that different levels of performance exist among freshman entering university agriculture programs. To accommodate the superior and the deficient student in the same program, use advance placement for superior students and let deficient students use an audio-tutorial center to catch-up outside of regular class. The junior college system is helping to reduce deficiencies in incoming students.

Agriculture students, in particular, seem to have low proficiency levels in chemistry, mathematics and physics at graduation from high school. It was felt that 4 years of mathematics in high school should be a minimum, and that high school physics would be more valuable than the chemistry as now taught.

Beginning chemistry courses in the university should prepare the agriculture student promptly for biochemistry. The courses should not be terminal. The group preferred that inorganic chemistry be given in the first year, biorganalytical chemistry the second year. This title has been adopted by UCLA in designating a set of courses developed by them; the basic philosophy of this course series is that analysis is part of almost all chemistry and that the rigor and precision of analytical thinking can be taught just as thoroughly by application to problems in organic and biochemistry as to these in the familiar inorganic systems. Every college of agriculture should work with the chemistry department so each knows what the other needs and expects.

Higher levels of mathematics are increasingly needed in agriculture courses. High school graduates should be ready for calculus as a first course in the university. Calculus is needed by forestry, agricultural mechanics, agricultural economics, agricultural science, graduate students and others.

Statistics is particularly necessary in courses in genetics, ecology, growth systems, agronomy, nutrition and animal breeding. Training in computer science is desirable to the extent that the student realizes the capabilities, the limitations and influences of the computer on modern agriculture. In order to encourage students to take more mathematics, teachers must use mathematics in their courses and require that the students do so.

The group agreed that a general survey course "about" physics is needed by all agriculture students, except for a few highly specialized ones; the latter should take courses "in" physics. Agricultural staffs need to work with the physics departments to develop suitable courses.

The following summary statements were formulated:

• There is great need for remedial work because students are not adequately prepared in mathematics, physics, and chemistry in high school.

- Chemistry deserves prompt attention because it is well established in the curriculum; mathematics and physics need to be considered on a similar level.
- Chemistry courses now taught do not altogether meet the needs of agriculture students.
- More mathematics and higher level mathematics is needed by most agriculture students.
- All agriculture students need physics, either as a survey type course or as in-depth theory courses.
- Consolidation and removal of duplication in courses and curricula is the key to giving students expanded knowledge in a four-year time span.
- There must be planning sessions among the involved departments, so that each knows what the other needs.

Comment 2

Knowledge of mathematics and physics is a part of being educated, and exposure to these two disciplines is an effective guard against obsolescence. Education of science-oriented students must prepare them to handle problems dealing in quantitative tersm. Moreover, principles of mathematics and physics have long-lasting application in many fields. Hence, although one cannot predict the specific courses that will be necessary 10 to 15 years hence, the principles of mathematics and physics always will be pertinent.

More specific reasons for improved proficiency in mathematics and physics follow:

- Advanced ideas and methods from the mathematical and physical sciences are already to be found in graduate courses in research and in the literature and practical activities in the field.
- Greater literacy of students and faculty in mathematics and physics would render the teaching of many key subjects more efficient.
- Increased mathematical competence will enable students to take full advantage of computers.
- Because of their nature, mathematics and physics are indispensable to the understanding and development of theory and practice in many fields—in 10 to 15 years the forefront of activity will require even more sophisticated mathematics and physics.
- Concepts and techniques resting on college-level mathematics and physics are already being introduced into upper-division sub-

stantive and supporting courses even though many students lack sufficient background.

TOPIC I

What level and amount of chemistry, physics and mathematics should undergraduates in agriculture and natural resources have? How should this vary by major field and by option (e.g., business, technology, science)?

Comment 1

The group reached the following conclusions:

- No course in chemistry, physics or mathematics for agriculture and natural resources students should be terminal.

A progressive program in chemistry should be offered that would provide a rapid integration of the subdisciplines including organic chemistry, biochemistry, and perhaps some physical chemistry. The program would vary with the major emphasis, i.e., business majors would take 6–8 semester hours, technologists would take 9–12 hours, and science majors would take 15–20 hours.

- A graduated program in mathematics, beginning with calculus, would provide 9–11 hours for business or education majors, 15–18 hours for technology majors, and 19–22 hours for science majors.

- A special one-year physics program is suggested that should emphasize fluid mechanics, thermodynamics, and other factors of special importance to those in agriculture, natural resources, and other biosciences. The course would not be inferior to that offered physics majors, but would have a different emphasis.

Comment 2

The group generally agreed that the following courses taken from the Report of the CEANAR Committee on Mathematics be recommended to colleges of agriculture and natural resources for adoption and implementation.

Course Name (Semester hours)	Recommended for curricula in		
	Education	Technology	Science
Introductory Calculus (3–4)	x	x	x
Multivariable Calculus (3–4)	–	x	x
Probability (3)	x	x	x
Linear Algebra (3–4)	–	–	x
Theory and Techniques of Calculus (3–4)	–	–	x
Statistical Inference (3)	–	x	x
Introduction to Computing (3–4)	x	x	–
Principles of Programming (1)	–	–	x
Total hours	9–11	15–18	19–22

To the question, can mathematics be taught with a more applied and less theoretical approach, the answer seems to be that people are asking for so much to be included in mathematics courses today there is little time left for application. The group felt that applications should be taught in the biological sciences courses, and that if the biological science teachers used mathematics in teaching their courses the students would see and understand the application. True enough, in time there might be courses called biomathematics, taught by biologists, just as there has been an increase in biochemistry. It seemed to be the opinion of those associated closely with forestry curricula that the mathematics listed under the column headed science would be desirable for forestry majors.

While not discussed in detail, there was some feeling that many agricultural students—and those unable to handle the minimum of 9 semester hours of mathematics—would be better served by a two-year program as technicians than in a four-year B.S. program. There was not much enthusiasm for telescoping mathematics subject matter into fewer courses and credits for science orientated students.

It should be pointed out that the 19-22 semester hours suggested for science students and the 15-18 hours suggested for technology students include those hours that now may be classified as statistics, and that the program is designed not for tomorrow but for ten years hence. Indeed, it probably would take ten years to accomplish even if it were decided now to do so.

As for chemistry, a new one-year course was thought to be an adequate minimum for all students in agriculture and natural resources. The group felt that chemistry and calculus should be taken concurrently. The second-year course suggested by the CEANAR Chemistry Committee was considered favorably but it was felt that some students in the science majors may well need additional courses.

The group endorsed the principles presented in the report of the CEANAR Committee on Physics, which suggests a minimum of one year of physics—not a watered-down course, but one with more emphasis on fluids and dynamics and less on mechanics than is usually included in freshman physics. The course should not be a terminal course but should permit students to take additional physics courses without penalty.

Comment 3

The group concurred with the recommendations of the Mathematics Committee (see table in Comment 2, above).

It recommended that the technology and science curricula list a one-year (8–10 hours) nonterminal course in physics of substantial content and utility. For other curricula, such as agricultural education and business orientated programs, the group recommended one year of physical sciences—not ordinarily a regular physics course. Its objectives and content should be a quality and at a level comparable to that of other physics courses for physicists and engineers, but very differently oriented with respect to academic objectives and drawing upon very different examples.

Comment 4

The members of the chemistry group agreed with the minimum needs contained in the CEANAR Chemistry Committee Report for students oriented toward social sciences and management in agriculture, natural resources and food technology curriculums. These needs translate into about 10 semester credits in an integrated chemistry program. Portions or all of the program are also suggested for all non-science university students, to insure that they too may have truly liberalizing education.

B. S. terminal science-oriented students should take additional offerings in organic and applied biochemistry.

Those students showing aptitude for graduate work in the science related areas will need more chemistry. Their courses should be chosen from among those required of the various chemistry majors.

Specific recommendations are as follows:

- The use of placement or proficiency tests:
    - to facilitate transfer of students from junior colleges, technical colleges and like institutions.
    - to take cognizance of high school preparation.
    - to help motivate students by placing them in a chemistry program that is meaningful to them.
    - to aid in transferring from one chemistry program to another,
    - to encourage independent study.
- Effective advising is required to minimize loss of time by placing students in the proper sequence early in their careers, depending on interests, capabilities, and readiness.
- Further development of compilations of examples of the application of chemistry in agriculture, natural resources and food technology so that chemistry teachers may relate to student interests is important.
- Resource persons from agriculture, natural resources, and food technology might serve as leaders in chemistry discussion periods, to show relevance and to encourage interest in chemistry, or joint appointments could be explored.

TOPIC II

How can recommendations be implemented? What can faculty members in agriculture and natural resources do? What can faculty members in chemistry, physics, and mathematics do?

Comment 1

- Wide variation among student qualification will continue indefinitely. Thus, continued improvement in student counseling, placement testing, and related services is essential to help place students at program levels consistent with their qualification. Remedial programs will be essential. Much of this service might be provided within the junior college.
- Arbitrary prerequisites should be waived and students encouraged to enter programs at whatever levels their backgrounds dictate.

- Implementation must begin at a local level; efforts must be devoted to tactfully selling the concepts first to agriculture and natural resources personnel, then to those involved in other biological disciplines, and finally to the chemistry, physics, and mathematics personnel.
- Resource materials and problem sets would aid curricula committee members.
- Financing of pilot teaching programs and publications of source materials would be helpful.
- Laboratories in the initial courses of a graduated program should be minimized now that improved laboratory work is offered at the high school.
- Instructors should continually up-grade their teaching methods and materials through incorporation of the latest physics, chemistry, and mathematics concepts. Course work designed specifically for faculty enrollment should be included on campuses to permit faculty to keep abreast of new developments and to facilitate improved dialogue.
- Team teaching among departments should be encouraged.
- Consolidation or elimination of some technical courses should be effected, to allow for an expanded program emphasizing fundamental principles, although a five-year program may eventually prove necessary.

Comment 2

- The faculties in agriculture and natural resources should be encouraged to:
  - improve their mathematical and physics appreciation and ability through course-work, post-doctoral, and sabbatic study and special summer-sessions.
  - use mathematics and physics more effectively in their courses, and prepare appropriate teaching materials.
- Cooperative programs should be developed by agriculture and natural resource faculties to encourage some of the faculty in mathematics and physics to:
  - improve their knowledge of mathematics and physics usage in agriculture and the natural resources.
  - provide sections of mathematics and physics courses in which illustrations and problems are drawn from agriculture and the natural resources and its supporting fields.

- jointly prepare appropriate teaching materials.
- Funds will be needed:
- for post-doctoral appointment and sabbatic leaves.
- for summer sessions devoted to mathematics in agriculture and natural resources that will be attended by faculty members.
- for individual or team preparation of texts and reference materials that can be used in mathematics and physics courses and in courses in agriculture and the natural resources.

● Through appropriate means all students in agriculture and natural resources should be encouraged to take physical science courses in high school so that, as soon as feasible, students entering four-year curricula in agriculture and the natural resources can be required to have had four years of college-preparatory mathematics.

● At least two pilot projects should be encouraged to develop a two-year sequence of courses that would integrate mathematics, chemistry, physics and biological science into one package.

APPENDIX F

Social Sciences

TOPIC I

What are the broad areas of social science and humanities on which agricultural and natural resources technology is dependent, and to which all students in agriculture and natural resources should be exposed? What broad areas of social science and humanities are essential to the student's liberal education?

Comment

The needs for social sciences and humanities in agriculture and natural resources are no different than in other areas. Core programs in social sciences and humanities at institutions of higher education seem to meet these needs.

The group felt that a minimum of 15% of the curriculum requirements should be in social sciences, with at least 7% (three courses) in one discipline and the remainder at least three other areas. Further recommendations were that a minimum of 15% of the curriculum requirements should be in the humanities, with at least 7% (three courses) in one discipline and the remainder at least three other areas.

TOPIC II

In what ways are the social sciences and humanities needs different for the various majors in the college of agriculture? In various options: business, technology, and science?

Comment

There are special needs in the social sciences and humanities in several areas of agriculture and natural resources depending on the major. These needs should be met on an individual student basis. Curricula should allow sufficient flexibility to meet these needs.

TOPIC III

What can be done to develop in students of agriculture and natural resources a greater consciousness of the relevance of social sciences and humanities to agriculture and natural resources?

Comment

Students can be made more aware of the relevance of social sciences and humanities through their professional course instructors. However, many students are more knowledgeable about the relevance of social sciences and humanities than are some professors. The group did not subscribe to the sourcebook concept, because it tends to shift responsibility away from the college where it belongs. Also, it is unrealistic at many multi-universities. To revise course contents regularly and keep them current will aid in maintaining relevance, and will make courses more palatable to students.

TOPIC IV

How can the recommendations be implemented?

Comment 1

The strong recommendation for heavy emphasis in social sciences and humanities is not unrealistic. Given most college and university core requirements, only two or three more courses need to be added to the agriculture and natural resources curricula to meet the recommended requirements.

A more serious problem of implementation is the desire to see that individual student needs are met more adequately. Several alternative routes are possible. Core requirements may need to be modified to allow more diversity. In some curricula additional room for more social sciences and humanities will have to be made for individuals who are oriented toward business-production type areas. This can and should be done at the expense of current mathematics, physics or professional course requirements.

Reviewed against the total agricultural curriculum, the recommendations made here are minimal. Without continuing emphasis and expansion in social science, our students will not be prepared to deal effectively with the dynamic, social and institutional forces influencing and directing agriculture and natural resources policies and actions.

The group felt strongly that increased course requirements and flexibility are a necessary but not a sufficient effort in social sciences and humanities. There is also a pressing need for incorporating principles of social sciences and humanities into professionally oriented subject matter areas. Only in this way will it be possible to insure that educational offerings and requirements are timely and relevant to current trends in our society. Responsibility for implementation of this recommendation rests with the individual faculty member. Compliance will be enforced by deans and students.

To provide for diverse student needs within individual curricula, more emphasis can be given to currently available courses in social sciences and humanities under such headings as economics, sociology, anthropology, georgraphy, political science, western civilization, music, art appreciation. These subjects can be identified as areas from which students must elect a specified minimum number of courses, specializing in at least one.

Comment 2

The group emphasized communication at individual campuses between biological departmental staff and students and the staff and students from the agricultural college. Specific suggestions include:

- Interdepartmental seminars
- Luncheons

- Joint appointments
- More time with people in other disciplines
- Curriculum committee discussion
- Cross listing of courses
- Interdisciplinary research

A sourcebook on a regional or local basis was recognized as one tool to provide reference sources and applications; committee members emphasized the limitation to realistic utilization of a sourcebook. Members felt that staff discussion, and familarity with course syallabi, provide a better means of identification of curricular needs. The need for proper motivation and faculty enthusiasm was recognized.

The group recognized that advisors must be aware of the content and importance of courses included in general education.

APPENDIX G

Core Curriculum

Purpose or goal for education in agriculture and natural resources:

- Attain professional and technical competence in an area of agriculture that will provide a base for professional employment.
- Provide a broad general education.
- Provide sound foundations in the fundamental sciences supporting agriculture as a basis for continued learning and maximum service.

The group recognized the importance of biological sciences, social sciences, humanities, communications, physical sciences (including mathematics) and professional agriculture and natural resources training in achieving the above-stated goals.

With the framework of courses available and the 125-130 credit framework now or in the foreseeable future available, it is necessary to choose those most appropriate to the needs of professional agriculture. Unintentional redundancy should be eliminated. It should be possible to bring needed knowledge and facts into sharp focus at specific points in the overall program and thereby save repetition in several areas.

Agriculture must demand the same support as is accorded other elements of the university in meeting the needs of agriculture, biology and prehealth programs. The large numbers of students currently enrolled in agriculture, biology and prehealth sciences will provide a strong base for winning this support.

Communication between the faculties of agriculture and biology and the faculties of supporting areas must be developed in order to implement desirable cooperation and constructive, creative changes.

Basic curriculum (minimum) (130 semester credits), as follows:

Core	Credits	Courses
English, speech, communications	12	4
Principles of Economics and Business	6	2
Chemistry (including introductions to organic chemistry and biochemistry)	10	2
Mathematics and Physics[a]	15–18	4–6
Biology	8	2
Humanities and Social Sciences (other than Principles of Economics and Agricultural Economics)	18	6
Professional Agriculture	24	6–8
Subtotal	85–88	
Elective (not more than 20 credits in professional agriculture courses)	42–45	
Total (approx)	130	

[a]To include elementary analysis, elementary calculus, elementary statistics, and probability and elementary computer science. Traditional courses in these areas must be restructured to come within 15 credits.

APPENDIX H

Organizers, Sponsors, and Speakers

WESTERN REGION CONFERENCE

University of Nevada, Reno   February 22–23, 1968

Conference Sponsors

Commission on Education in Agriculture and Natural Resources
Western Region, Resident Instruction Section, Division of Agriculture, National
   Association of State Universities and Land-Grant Colleges
Western Region, National Association of Colleges and Teachers of Agriculture

Steering Committee

Darrel S. Metcalfe, *Chairman* (CEANAR representative), University of Arizona
Lloyd Dowler (NACTA representative), Fresno State College
Charles H. Seufferle (Local representative), University of Nevada
E. C. Stevenson (NASULGC representative), Oregon State University

NORTHEAST REGION CONFERENCE

Hotel America, Hartford, Conn.　May 2-3, 1968

Conference Sponsors

Commission on Education in Agriculture and Natural Resources, (CEANAR)
Northeast Region, Deans and Directors of Resident Instruction in Agriculture,
　　National Association of State Universities and Land-Grant Colleges
　　(NASULGC)
Northeast Region, National Association of Colleges and Teachers of Agriculture
　　(NACTA)

Conference Steering Committee

O. J. Burger, Fresno State College (Representing NASULGC)
Burton W. DeVeau, Ohio University (Representing NACTA)
E. J. Kersting, University of Connecticut (Representing the University of
　　Connecticut—host institution)
Russell E. Larson, *Chairman*, The Pennsylvania State University (Representing
　　CEANAR)

NORTH CENTRAL REGION CONFERENCE

University of Wisconsin, Madison　March 13-14, 1969

Conference Sponsors

Commission on Education in Agriculture and Natural Resources (CEANAR)
North Central Region Deans and Directors of Resident Instruction in Agriculture
National Association of State Universities and Land-Grant Colleges (NASULGC)
Northeast Region of the National Association of Colleges and Teachers of
　　Agriculture (NACTA)

Steering Committee

Carroll V. Hess, *Chairman*, Kansas State University
George Sledge, University of Wisconsin
Eugene Wood, Southern Illinois University
Lee Baker, Western Michigan University

SOUTHERN REGION CONFERENCE

University of Georgia, Athens   October 21–22, 1969

Conference Sponsors

Commission on Education in Agriculture and Natural Resources
Southern Division, Resident Instruction Committee on Policy, National Association of State Universities and Land-Grant Colleges
Southern Region, National Association of Colleges and Teachers of Agriculture

Steering Committee

Hal B. Barker, Louisiana Tech University, *Chairman*
Murray Brown, Sam Houston State College, Huntsville, Texas
Glenn Hardy, University of Arkansas
Fred Manley, North Carolina Department of Community Colleges
Robert S. Wheeler, University of Georgia

SPEAKERS

Duane C. Acker, Dean, College of Agriculture and Biological Sciences, South Dakota State University, Brookings, S.D. 57006
David G. Barry, Vice President and Provost, The Evergreen State College, Olympia, Washington 98501
James S. Bethel, College of Forest Resources, University of Washington, Seattle, Washington 98105
F. Yates Borden, School of Forest Resources, The Pennsylvania State University, University Park, Penna. 16802
Robert H. Burris, Department of Biochemistry, University of Wisconsin, Madison, Wisconsin 53706
Burton W. DeVeau, College of Business Administration, Ohio University, Athens, Ohio 45701
Thomas W. Dowe, Dean, College of Agriculture and Home Economics, University of Vermont, Burlington, Vermont 05401
J. Clyde Driggers, President, Abraham Baldwin Agricultural College, Tifton, Georgia 31794
George A. Gries, Dean, College of Arts and Sciences, Oklahoma State University, Stillwater, Oklahoma 74074
Carroll V. Hess, Dean, College of Agriculture, Kansas State University, Manhattan, Kansas 66502
John F. Hosner, Director, Division of Forestry and Wildlife, College of Agriculture, Virginia Polytechnic Institute, Blacksburg, Va. 24061